高等学校教材

聚合物降解与稳定：基础与实践
（英文版）
Polymer Degradation and Stabilization: Fundamental and Practice

闫毅　姚东东　孔杰　颜静　编著

西北工业大学出版社

西　安

【内容简介】 本书共7个部分，内容包括绪论，不同类型的聚合物降解，如热降解、热氧化降解、光降解、光氧化降解、臭氧降解及微生物降解，表征聚合物降解的不同方法，在不同降解条件下稳定聚合物的一般方法，一些典型通用聚合物(如聚苯乙烯、含氟聚合物、聚乙烯醇、基于甲基丙烯酸酯/丙烯酸酯的聚合物和聚二烯烃等)的降解和稳定。

本书可作为高分子化学与物理、高分子工程与材料及应用化学专业本科生和研究生的教材，也可供从事高分子化学相关领域的科研人员参考。

图书在版编目(CIP)数据

聚合物降解与稳定：基础与实践 ＝ Polymer Degradation and Stabilization：Fundamental and Practice：英文 / 闫毅等编著. — 西安：西北工业大学出版社，2023.12
ISBN 978 - 7 - 5612 - 9082 - 8

Ⅰ. ①聚… Ⅱ. ①闫… Ⅲ. ①高聚物-降解-英文②高聚物-稳定化处理-英文 Ⅳ. ①TQ316

中国国家版本馆 CIP 数据核字(2023)第 209966 号

JUHEWU JIANGJIE YU WENDING：JICHU YU SHIJIAN
聚 合 物 降 解 与 稳 定：基 础 与 实 践
闫毅 姚东东 孔杰 颜静 编著

| 责任编辑：胡莉巾 倪瑞娜 | 策划编辑：杨 军 |
| 责任校对：杨 兰 | 装帧设计：李 飞 |

出版发行：西北工业大学出版社
通信地址：西安市友谊西路 127 号 邮编：710072
电　　话：(029)88491757，88493844
网　　址：www.nwpup.com
印　刷　者：兴平市博闻印务有限公司
开　　本：787 mm×1 092 mm　　1/16
印　　张：12.25
字　　数：321 千字
版　　次：2023 年 12 月第 1 版　2023 年 12 月第 1 次印刷
定　　价：ISBN 978 - 7 - 5612 - 9082 - 8
定　　价：49.00 元

如有印装问题请与出版社联系调换

前　言

当我们谈到聚合物降解时,你脑海中的第一个想法是什么?在回答这个问题之前,让我们思考以下问题:

(1)什么是化学和化学反应?如何合成新的化合物?

(2)你怎么知道什么食物是可以吃的?我们的身体是如何消化食物的?

(3)你知道白色污染吗?你对处理白色污染有什么建议吗?你对垃圾分类有什么建议吗?

很显然,这些问题都与聚合物降解有关,如聚合物降解过程中的化学过程和可生物降解的聚合物等。根据国际纯粹与应用化学联合会(International Union of Pure and Applied Chemistry,IUPAC)的定义,聚合物降解是指聚合物材料中的化学变化,这些变化通常会导致材料在使用中的性能发生不希望的改变。大多数情况下聚合物(如乙烯基聚合物、聚酰胺)降解伴随着分子量的降低。而在某些情况下,如主链上含有芳环的聚合物降解意味着其化学结构的变化,如交联。一般来说,降解会导致材料有用性能的损失或劣化。然而,在某些情况下,聚合物降解变得非常有用。例如,生物降解(通过生物活性降解)可能会将聚合物转变为具有理想性质的环境友好物质。

因此,在本书中,我们试图涵盖以下基本理论和实践:

(1)为什么我们需要研究聚合物的降解?

(2)聚合物如何降解?在这个过程中发生了什么样的化学过程?是什么样的机制?

(3)聚合物降解在生产实际中的应用。

(4)聚合物降解的实验研究。

本书的目的是让学生了解聚合物降解及其应用,重点培养学生分析和解决问题的能力。我们希望学生能够通过本书:

(1)了解聚合物降解的基本机理。

(2)了解聚合物降解过程中的基本反应。

(3)了解聚合物降解的应用。

(4)根据实际需求,设计特定聚合物降解实验。

本书的编写是由闫毅教授、姚东东副教授、孔杰教授和颜静副研究员共同完成的。其中,闫毅教授编写第 1~3 章,姚东东副教授编写第 7 章,孔杰教授编写第 4~5 章,颜静副研究员编写第 6 章。闫毅教授完成最后统稿。

感谢西北工业大学化学与化工学院郑亚萍教授和陈立新教授在本书编写过程中提出的宝贵意见。也非常感谢杨军在本书的编写和出版过程中给予的帮助。

由于笔者水平有限,书中难免有疏漏之处,恳请读者批评指正。

编著者
2022 年 9 月

Preface

When we talk about polymer degradation, what's the first idea in your mind? Let's think about the following questions:

(1) What's chemistry and chemical reaction? How to synthesize new compounds?

(2) How can you tell what kind of food is eatable? How does our body digest food?

(3) Do you know anything about white pollution? Do you have any proposals to deal with white pollution? Do you have any suggestions about garbage classification?

Apparently, these questions are related to polymer degradation, such as the chemical processes during polymer degradation, biodegradable polymer, etc. According to International Union of Pure and Applied Chemistry (IUPAC), polymer degradation is defined as the chemical changes in a polymeric material that usually results in undesirable changes in the inuse properties of the material. In most cases (such as vinyl polymers, polyamides), degradation is accompanied by a decrease in molar mass. While in some cases, such as polymers with aromatic rings in the main chain, degradation means changes in the chemical structure, such as crosslink. Usually, degradation results in the loss of (or deterioration in) useful properties of the material. However, in some cases polymer degradation becomes very useful. For example, polymers may change into environmentally acceptable substances with desirable properties in biodegradation (degradation by biological activity).

Therefore, in this book, we are trying to cover both the basic theory and practice as following:

(1) Why do we need to study polymer degradation?

(2) How can polymers be degraded? What kind of chemical process will happen in this process and what kind of mechanism is it?

(3) Polymer degradation in practical applications.

(4) Laboratory practice of polymer degradation.

The aim of this book is to expose students to the field of polymer degradation. The development of analytical and problem-solving skills will also be emphasized. We wish the readers of this book will be able to:

(1) Understand basic mechanism of polymer degradation.

(2) Understand the basic reactions during polymer degradation.

(3) Understand the applications of polymer degradation.

(4) Design simple experiments for specific polymer degradation.

This book was written by Professor Yan Yi, Associate Professor Yao Dongdong, Professor Kong Jie and Associate Researcher Yan Jing. Amon them, Professor Yan Yi wrote Chapters 1 – 3, Associate Professor Yao Dongdong wrote Chapter 7, Professor Kong Jie wrote Chapters 4 – 5, Associate Researcher Yan Jing wrote Chapter 6. Professor Yan Yi finished the final draft.

The authors thank Prof. Zheng Yaping and Prof. Chen Lixin from School of Chemistry and Chemical Engineering of Northwestern Polytechnical University for their kind advices during the process of writing this book. We also thank Mr. Yang Jun for his help during the preparation and publication of this book.

Because of the limited knowledge of the authors, this book has to avoid inaccuracies, welcome any suggestions and criticism.

Authors

September 2022

Contents

Chapter 1 Introduction ··· 1

1.1 Introduction to polymer degradation ··· 1
1.2 Phenomena and causes of polymer degradation ····································· 1
1.3 Necessity and possibility of polymer stabilization ································· 11
1.4 Recycling of polymers and biodegradable polymers ······························· 16
Questions ·· 18
Further reading Ⅰ: polymer degradation ·· 18
Further reading Ⅱ: hydrolysis of condensation polymers ······························ 19
Further reading Ⅲ: hydrolysis of polyesters ·· 21
Further reading Ⅳ: biodegradable polymers utilized in food packaging ·········· 23

Chapter 2 Thermal degradation and stabilization ··································· 26

2.1 Thermal degradation ··· 26
2.2 Stabilization of thermal degradation ··· 34
2.3 Thermal-oxidative degradation ··· 36
2.4 Stabilization of thermal-oxidative degradation ····································· 42
Questions ·· 50
Further reading Ⅰ: thermal degradation ·· 51
Further reading Ⅱ: auto-accelerated oxidation of plastics ····························· 53
Further reading Ⅲ: effects of peroxides and antioxidants on auto-oxidation of plastics ········· 54
Further reading Ⅳ: thermal-oxidative degradation of rubbers ························ 55
Further reading Ⅴ: general mechanism of thermal degradation ····················· 56
Further reading Ⅵ: antioxidants ·· 58
Further reading Ⅶ: sterically hindered phenolic antioxidants ························ 59

Chapter 3 Photo degradation and stabilization ····································· 61

3.1 Photo-degradation ·· 61
3.2 Photo-oxidative degradation ·· 68
3.3 Stabilization of photo degradation and photo-oxidative degradation ········· 74
3.4 Photodegradable polymers ··· 78
Questions ·· 78

Further reading Ⅰ: photo degradation of polymers ……………………………………… 79
Further reading Ⅱ: hindered amine light stabilizers (HALS) ………………………… 80
Further reading Ⅲ: metal deactivators and UV-light absorbers …………………… 82
Further reading Ⅳ: free radical photoinitiators ……………………………………… 83
Further reading Ⅴ: radiolytic degradation ……………………………………………… 85

Chapter 4　Ozone degradation and stabilization ……………………………………… 93

4.1　Ozone degradation ……………………………………………………………………… 93
4.2　Stabilization of ozone degradation ………………………………………………… 95
Further reading Ⅰ: ozonation reactions and ozone cracking in elastomers ……… 95
Further reading Ⅱ: antiozonants *para*-phenylenediamines ………………………… 96

Chapter 5　Microbiological degradation and stabilization ………………………… 98

5.1　Basics for biodegradation …………………………………………………………… 98
5.2　Stabilization of biodegradation ……………………………………………………… 100

Chapter 6　Characterization of polymer degradation ……………………………… 101

6.1　Mechanical tests ………………………………………………………………………… 102
6.2　Gel permeation chromatography …………………………………………………… 104
6.3　Fourier transform infrared spectroscopy ………………………………………… 105
6.4　Magnetic resonance spectroscopy ………………………………………………… 107
6.5　Oxygen uptake …………………………………………………………………………… 109
6.6　Chemiluminescence …………………………………………………………………… 110
Further reading Ⅰ: thermogravimetric analysis of polymers ……………………… 111

Chapter 7　Degradation and stabilization of different class of commodity polymers …… 114

7.1　Degradation and stabilization of polyolefins ……………………………………… 114
7.2　Degradation and stabilization of polyvinyl chloride and other chloride-containing polymers ……………………………………………………………………………… 130
7.3　Degradation and stabilization of polystyrene and styrene copolymers ……… 140
7.4　Degradation and stabilization of fluoropolymers ………………………………… 152
7.5　Degradation and stabilization of other vinyl polymers ………………………… 154
7.6　Degradation and stabilization of acrylate and methacrylate polymers ……… 162
7.7　Degradation and stabilization of polydiene polymers …………………………… 169
Further reading Ⅰ: thermal oxidative degradation of polypropylene …………… 175
Further reading Ⅱ: PVC heat stabilizers ……………………………………………… 178

Appendix Ⅰ　Typical wordlist used in this book ……………………………………… 179
Appendix Ⅱ　Chemical structure, name and abbreviation of typical polymers …… 182
Appendix Ⅲ　Typical parameters for the characteristic viscosity-molecular weight relationship of typical polymers ……………………………………… 186
References ……………………………………………………………………………………… 187

Chapter 1　Introduction

1.1　Introduction to polymer degradation

The concept of "macromolecular chemistry" has been introduced by Hermann Staudinger for a century. Due to their unique properties in electronics, optics, as well as their processabilities and mechanical properties, polymers and related products have been widely used in both daily life and industry. However, during the process of storage and application of polymers, there are some problems obstructing the progress of polymeric materials, such as the chemical structure of polymers may change, the mechanical properties of polymers become worse, the durabilities of polymeric products become worse, etc. All of these problems are related to so called "polymer degradation".

Therefore, one of the key researches focuses of polymer degradation is to understand the basic chemical processes and mechanisms during the degradation of polymers, find possible solutions to improve or stabilize the properties of polymeric products, and prolong the lifetime of these products. Besides the study of these scientific questions, another research topic of polymer degradation focuses on green chemistry of polymers, such as solutions to white pollution, biodegradable polymers, biomass-based polymers, etc.

To understand the details and unique properties of polymer degradation, this book will cover the following contents:

(1) Basic principles of polymer degradation and stabilization (**Chapter 1**).

(2) Thermal (**Chapter 2**)/photo (**Chapter 3**)/ozone (**Chapter 4**)/microbiological (**Chapter 5**) degradation and stabilization.

(3) Characterizations of polymer degradation (**Chapter 6**).

(4) Degradation of some classical polymers (**Chapter 7**).

1.2　Phenomena and causes of polymer degradation

1.2.1　Phenomena of polymer degradation

Polymer degradation is a world-wide problem during the applications of natural polymers and synthetic polymers as well as their products.

Here are some examples of polymer degradation:

(1) After long period applications of fiber and cotton products, these clothes become yellow and brittle, this is the phenomena of degradation of natural polymers.

(2) Synthetic polymers such as rubbers (tires, gloves) become sticky, harden, cracking after long period of applications or after use in special conditions.

(3) Thermoset resin, such as phenolic resin, may become yellow or even dark if processing under higher temperature or longer time, this is polymer degradation during processing.

(4) The transparency of polystyrene will be decreased after certain time of storage and usage, some morphology damage or even cracking will happen, this is degradation phenomena of thermoset polymers.

As mentioned above, polymer degradation is associated with polymer itself, actually degradation is accompanied with polymer as long as the polymer is synthesized or the polymeric product is prepared. Generally, the characteristics of polymer degradation at molecular scale is the decrease of molecular weight and the change of molecular weight distributions. At the very beginning, the change of molecular weight cannot be visualized; while dramatic change in mechanical and physical properties will be observed after certain degree of degradation. As shown in Fig. 1.1, the molecular weight of polypropylene (PP) decreases with the ultraviolet (UV) irradiation time. One can draw the following conclusions:

(1) At the very beginning of irradiation, the molecular weight decreases dramatically, which shows a linear relationship with the irradiation time. However, the mechanical and physical properties of PP can be maintained at this stage.

(2) After long time of irradiation, the films become brittle, small cracks can be observed on the surface of the PP films, and the mechanical properties of these films become worse.

It should be mentioned that some active centers generated during the degradation may induce crosslink, which can increase the molecular weight. However, chain breaking is dominating during the process of polymer degradation.

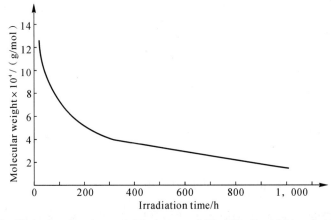

Fig.1.1 Plot of molecular weight change of PP with the UV irradiation time

Normally, polymer degradation always brings disadvantages to polymeric products, however, one can also make use of polymer degradation.

Over 95%(wt.) of polymers are synthesized from eight vinyl monomers, and most of the remainders are from another couple of dozen monomers with diverse chemical structures. Despite this relative simple chemistry, the problems of degradation and stabilization of synthetic polymers are exceedingly complex. This complexity stems from the structural organization of plastic materials, which occurs at different levels:

(1) At the molecular level, most commercial plastics are mixtures of macromolecules with diverse substances, which can be traces of solvent or catalyst residues, pigments, fillers, and a variety of additives.

(2) The macromolecules themselves can be blends of homopolymers or copolymers.

(3) As to the differences in small organic molecules, even a simple homopolymer presents a diversity in chain length and configuration. The presence of chain branching, tacticity, monomer linking, or chain defects may all have important consequences for the material's stability.

(4) Molecular chains can arrange themselves into amorphous, oriented, and crystalline domains of different sizes, shapes, and distributions. This intricate morphology is globally known as supramolecular organization.

Studying degradation requires not only identification of the diverse chemical reactions which induce structural changes in the polymer, but also the interactions between the numerous chemical species initially present in the sample or formed in the course of degradation. Interdependence of the change in material morphology that occurs upon aging, and its influence on the rate of degradation, should be considered additionally.

1.2.2 Causes of polymer degradation

Basically, there are two kinds of causes for polymer degradation: intrinsic cause and extrinsic cause. Intrinsic cause is related to the polymer itself, while extrinsic cause is about the external environment and condition of polymers. Polymer degradation is a complicated process, which always involves both intrinsic and extrinsic causes. In order to stabilize a polymer, both intrinsic and extrinsic causes should be considered.

The composition and chain structure of polymers, their corresponding aggregated state and impurities introduced during polymer synthesis and processing are intrinsic causes, while extrinsic causes include light, heat, oxygen, chemicals, mechanical force, microbial, etc.

1. Intrinsic causes

(1) Composition and chain structure of polymers. Polymers are monomers connected through covalent bond. Polymer degradation involves the breakage of chemical bonds. The

less the bonding energy of the chemical bond, the easier the polymer degradation. For example, polyethylene(PE) and polytetrafluoroethylene (PTFE) have the similar structure (see Fig. 1.2), however, their degradation behavior is completely different. PTFE shows excellent property and great stability toward chemicals, heat, light, etc. Untreated polyethylene film with 0.1 mm thickness display obvious degradation after 2 - 3 months, while in the case of PTFE film even after 75 months no obvious degradation can be observed.

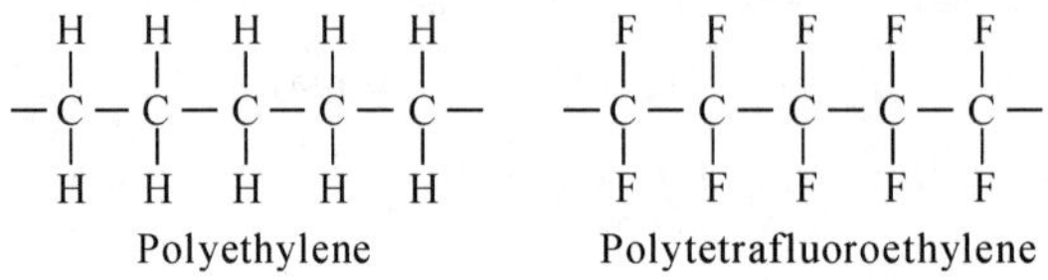

$$-\underset{H}{\overset{H}{C}}-\underset{H}{\overset{H}{C}}-\underset{H}{\overset{H}{C}}-\underset{H}{\overset{H}{C}}-\underset{H}{\overset{H}{C}}- \qquad -\underset{F}{\overset{F}{C}}-\underset{F}{\overset{F}{C}}-\underset{F}{\overset{F}{C}}-\underset{F}{\overset{F}{C}}-\underset{F}{\overset{F}{C}}-$$

Polyethylene   Polytetrafluoroethylene

Fig. 1.2 Chemical structure of polyethylene and polytetrafluoroethylene

From the view of polymeric structure, the only difference between these two polymers is the fluorine and hydrogen. Because of the difference of the electronegativity of these two atoms, the bond energy of C—F and C—H is different. Normally the bond energy of C—F is 500 kJ/mol, while C—H is 410 kJ/mol, which may explain the stability difference of these two polymers.

Besides C and H, there are some other atoms and groups in polymer structure, such as unsaturated bond, hydroxyl, carboxylic, amide, ester, C—S in polysulfone, and so on. The bond energy of these bonds is normally weaker than C—C bond. Therefore, these bonds and groups are the weak point of the polymer structure, which are the active site for polymer degradation.

For polymers constructed from C—H, the architecture of the polymers also influences their degradation behavior. For example, linear PP can be oxidized more easily than linear polyethylene in practical application (see Fig. 1.3). From the study of oxidation of polymers, one can find that the reaction rate is dependent on the difficulty of proton abstract activity of the oxidant from polymers, while the rate of proton abstract is dependent on the kinds of C—H structure. The activity of C—H in polymer follows the same sequence as small molecule: the proton in the branched knot [see Fig. 1.3(a)] is much easier to be removed than the protons in methylene groups [see Fig. 1.3(b)]. Therefore, the proton on the methyl groups or chain end [see Fig. 1.3(c)] is more stable. This can explain the oxidation stability of PE and PP.

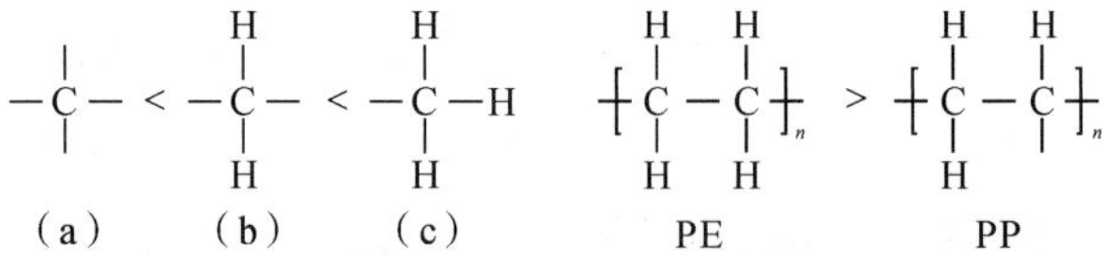

$$-\underset{\;}{\overset{\;}{C}}-\; < \; -\underset{H}{\overset{H}{C}}-\; < \; -\underset{H}{\overset{H}{C}}-H \qquad \left(\underset{H}{\overset{H}{C}}-\underset{H}{\overset{H}{C}}\right)_n \; > \; \left(\underset{H}{\overset{H}{C}}-\underset{\;}{\overset{H}{C}}\right)_n$$

(a)   (b)   (c)   PE   PP

Fig. 1.3 Stability sequence toward oxidation for small molecules and polymers

Chapter 1 Introduction

The difficulty of proton abstraction is a useful rule to determine the oxidation stability of polymers. There are also some other factors to be considered in practical application. For example, the structure of polystyrene (PS) and PP is quite similar, the only difference is methyl and phenyl group (see Fig. 1. 4). According to the above rule, PP should be more stable than PS due to the phenyl proton is more active than the proton in PP.

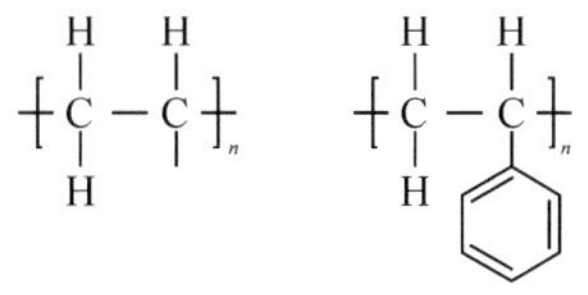

Fig. 1.4 Chemical structures of polypropylene and polystyrene

However, PS is more stable than PP toward oxidation in practical application. W. L. Hawkins and Hansen thought the stability difference is attributed to the shielding effect of the phenyl ring to the C—H. Dr. Hansen carried out a systematic study on the shielding effect by extending the link between polymer backbone and phenyl ring.

The shielding effect of phenyl group can be demonstrated by the data shown in Table 1. 1. Experimentally, by inserting one, two, ⋯ and four methylene groups into PS structure, the stability toward oxidation follows the sequence of 1>2>3>4.

Table 1. 1 Stability of polystyrene and its derivatives toward oxidation

Polymer	Chemical struture	Induction period/h	
		80 ℃	110 ℃
1	$+CH_2-CH(C_6H_5)-H+_n$	—	>10,000
2	$+CH_2-CH(CH_2-C_6H_5)+_n$	>10,000	1,900
3	$+CH_2-CH((CH_2)_2-C_6H_5)+_n$	500	30
4	$+CH_2-CH((CH_2)_4-C_6H_5)+_n$	200	13

(2) Molecular weight and distribution. The influence of molecular weight on degradation behavior is different for different polymers. In some cases, the stability increases with the increasing of molecular weight, such as polyvinyl chloride (PVC), polymethylmethacrylate (PMMA); while in some cases, the stability of polymer is not related to its molecular weight, such as polyisobutene (PIB). The influence of molecular weight is related to their irregular structures. If the increasing of molecular weight will decrease the content of irregular structure, the stability will be enhanced; vise versus.

The distribution of molecular weight definitely influences the stability of polymer. As shown in Fig. 1.5, polymer **a** shows a broad distribution, indicating large content of small molecular weight, which is easier to be decomposed. For polymer **b** with a narrow molecular weight distribution, it should show a better stability than polymer **a**.

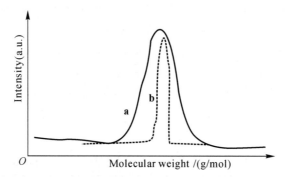

Fig. 1.5 Gel Permeciton Chromatography (GPC) traces for two polymers with different molecular weight distribution

(3) Degree of branching. The term degree of branching (DB) was introduced as a quantitative measure of the branching perfectness for hyperbranched polymers. Polymer with large degree of branching always has more weak points, which usually results in poor stability. For example, for low density polyethylene (LDPE), there are 8 – 10 branched chains per 1,000 carbon atoms, while there are less than 5 in the case of high density polyethylene (HDPE). As a result, the thermal stability of LDPE is worse than HDPE.

(4) Aggregation state. There are some aggregation states for polymers: crystalline state, amorphous state, orientation texture, rubbery state, supramolecular state, etc. Polymer degradation is also related to their aggregation states. In general, the density of non-crystalized materials is lower than crystalized materials. As a result, oxygen, water, and chemicals can easily penetrate non-crystalized materials and result in corresponding degradation. For example, branched PE is much easier to be decomposed than linear PE under the same condition. As shown in Fig. 1.6, the reaction and O_2 adsorption ability of crystalline linear PE is lower than that of branched PE.

Chapter 1 Introduction

It is generally considered that only the amorphous regions in semicrystalline polymers can participate in the degradation reactions. The chemical inertness of the crystalline phase originates from the close packing of the polymer chains (minimum free volume), from the rigidity of the crystal lattice, and from the lack of oxygen to initiate oxidative degradation. Only small species such as hydrogen atoms can penetrate the crystal lattice. Any macroradical which may be created is virtually trapped in the crystallites as an inert species. From the effects mentioned, it is obvious that degradation stability parallels with the degree of crystallinity. In some infrequent circumstances, it has been observed that increasing the degree of crystallinity in a polymer sample may increase the rate of photo-degradation. Closer scrutiny of the results has indicated that multiple light scattering by the crystallites, which increases the effective optical length, and hence the number of absorbed photons for degradation, is the proper explanation for this unusual behavior. In addition to the degree of crystallinity, the size of the crystallites can also influence the polymer stability through the dual action of chain orientation and the presence of tie molecules. A quenched LDPE, for instance, has a higher resistance to photo-oxidation than an annealed sample of the same degree of crys tallinity, as a result of smaller crystallite sizes and a greater number of tie molecules. In isotactic polypropylene (i-PP), a reverse trend has been observed when the rate of oxygen diffusion becomes the control step. It has been suggested that oxygen diffusion may depend on the type of amorphous phase, which is distinguished in PE as intralamellar, interlamellar, and interspherulitic.

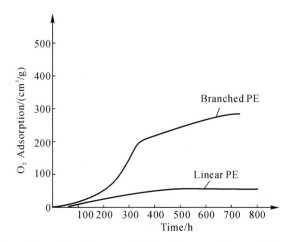

Fig.1.6 The adsorption diagram of O_2 for linear and branched polyethylene

Although it is well established that oxidative degradation of semicrystalline polymers initiates and spreads in the amorphous interphase without affecting the crystalline domains, it does change the density and crystalline content through a process known as chemicrystallization. The best documented account of this phenomenon is found in isotactic polypropylene. At the beginning of degradation, chain shortening from bond scission facilitates molecular rearrangement, leading to an increase in crystallinity. For long degradation times the crystallinity decreases, sometimes with a shift toward lower T_m, even if the degradation temperature is too low to allow for a change in molecular conformation.

(5) Impurity. Impurity can be introduced to polymer and polymeric product during both the synthesis and processing. There are different kinds of impurities, such as initiator, unreacted monomer, side product during polymerization; auxiliary I during the reaction (e. g. emulsifier, dispersant, stabilizer); auxiliary II during the processing and molding (e. g. dye, pigment, plasticizer, filler); other impurities, such as metal ions introduced from the reactor and processing instrument.

2. Extrinsic causes

(1) Heat, temperature, and oxygen. Most chemical reactions involve heat process. For polymers and polymeric products, heat may accelerate the movement of polymer chains and induce possible crosslink and degradation, as well as the decrease of physical and mechanical properties. During the breakage of chemical bond of polymer chains, it may generate some radicals. Such radicals may react with other species, such as radicals and impurities.

In presence of O_2, these radicals will react with O_2 to form peroxyl radicals, and involve in the further reaction with alkyl radicals or peroxyl radicals. If the radicals formed on the side chain, such radicals will also induce crosslink between these polymer chains.

In presence of O_2, the following reaction may happen:

$$\sim\sim CH_2-\cdot CH_2 + O_2 \longrightarrow \sim\sim CH_2-CH_2-O-O\cdot$$

Further reaction with each other or alkyl radicals may also happen:

$$\sim\sim CH_2-CH_2-O-O\cdot + \cdot O-O-CH_2-CH_2\sim\sim \longrightarrow \sim\sim CH_2-CH_2-O-O-CH_2-CH_2\sim\sim + O_2$$

$$\sim\sim CH_2-CH_2-O-O\cdot + \cdot CH_2-CH_2-CH_2\sim\sim \longrightarrow \sim\sim CH_2-CH_2-O-O-CH_2-CH_2-CH_2\sim\sim$$

The crosslink between polymer chains may happen as follows:

$$\sim\sim CH_2-\cdot CH-CH_2\sim\sim + \sim\sim CH_2-\cdot CH-CH_2\sim\sim \longrightarrow \begin{array}{c}\sim\sim CH_2-CH-CH_2\sim\sim\\|\\\sim\sim CH_2-CH-CH_2\sim\sim\end{array}$$

Heat-induced degradation may happen during polymer processing. For example, during the injection or extrusion processing of PVC, trace amount of O_2 may induce the degradation of PVC, and even carbonization.

As we know, the processing temperature of polymer should be lower than its glass transition temperature (T_g). Besides high temperature, low temperature may also induce polymer degradation. For example, rubber and corresponding product become brittle at low temperature.

(2) Light. Sunlight is one of the most important factors for polymer degradation. For most of the polymers and polymeric products used outside, the influence of sunlight on degradation should be considered. As shown in Fig. 1.7, sunlight covers the wavelength from 250 nm to 2,500 nm, including UV, visible, and infrared (IR) region.

1) UV: 150 - 400 nm, 5% of sunlight.

2) Visible: 400 – 800 nm, 40% of sunlight.
3) IR: 800 – 3,000 nm, 55% of sunlight.

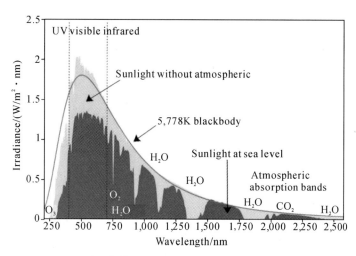

Fig. 1.7 The spectrum of sunlight

The energy of the light is dependent on wavelength. The smaller the wavelength, the higher the energy. Although UV region is only 5% of the sunlight, due to its short wavelength, the energy of UV light is high, which is strong enough to break chemical bonds. Table 1.2 gives the energy of light with different wavelength. Compare with the bond energy of the chemical bonds listed in Table 1.3, one can find that UV light can break most of the chemical bonds and the energy of visible light is also enough to break —O—O— bond.

Table 1.2 Energy of light with different wavelength

Light	Wavelength/nm	Energy/kJ
Microwave	$10^6 - 10^7$	$10^{-1} - 10^{-2}$
IR	$10^3 - 10^6$	$10^{-1} - 10^2$
Visible	800	147
	700	171
	600	201
	500	239
	400	299
UV	300	399
	200	599
	100	1,197
X-ray	10^{-1}	10^6
γ-ray	10^{-3}	10^8

Table 1.3 Bond energy of different chemical bonds

Chemical bond	Bond energy/(kJ/mol)
O—O	138.9
C—S	259.4
C—N	291.6
C—Cl	328.4
C—C	347.7
C—O	351.5
N—H	390.8
C—H	413.4
H—H	436.0
O—H	462.8
C=C	607.0

The interaction between light and polymer is different from the way heat interacts with polymer. Only when photonsare absorbed by polymer, the corresponding reactions may happen. It is very slow for photo degradation, because of the selectivity and intensity of photon absorption. Table 1.4 lists the sensitive wavelength of some classical polymers.

The energy of IR light is very weak. However, the absorption of IR will generate heat, which should have the same effect as thermal and thermal-oxidative degradation.

Table 1.4 Sensitive wavelength of some classical polymers

Polymer	Sensitive wavelength/nm
PC(polycarbonate)	285-305, 330-360
PE	300
PVC	320
PS	318.5
PP	300
PMMA	290-315
P(VC-VAc) [poly(vinyl chloride-vinyl acetate copolymer)]	327, 364
PET(polyethylene terephthalate)	325
POM(polyformaldehyde)	300-320

(3) Oxygen and ozone (O_3). Most of the polymers are used under the circumstance of atmosphere. The reactivity of O_2, CO_2, and other active components in air toward polymers

should be considered. As shown in Fig. 1.8, there is more than 20% O_2 in air. Oxidation, thermal-oxidative, and photo-oxidative degradation may happen in certain conditions. The oxidation activity of O_3 is much stronger than O_2, which will be introduced in **Chapter 4**. O_3 can react with the unsaturated bond in resin, which usually results in so called "ozone cracking".

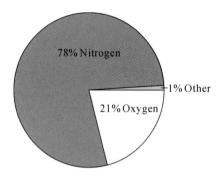

Fig.1.8 Composition of air

(4) Water. Water can penetrate polymers and react with the weak bonds, such as ester, amide, aldehyde group, and so on. The hydrolysis of these weak bonds will definitely result in polymer degradation. In some cases, the influence of water on polymer is reversible, for example, the mechanical change of nylon in presence and absence of water.

(5) Other factors. Other factors refer to the environment and external condition of polymeric products. Microorganism such as fungus and bacteria, and other organisms such as insect, rodent, and moth can also be potential influence factors for polymer degradation, which is called biodegradation.

During the synthesis and process of polymers, mechanical force cannot be avoided. For example, during the milling of rubber, strong shearing force may break the chemical bond and result in an obvious decrease of molecular weight. In some cases, mechanical degradation of polymers can also be used.

1.3 Necessity and possibility of polymer stabilization

Compare with inorganic and metallic materials, the life time of polymeric materials is very short. Since polymer degradation cannot be avoided, the goal of "Polymer Degradation" is to prolong the life time of polymer under the premise of suitable degradation and find green process for polymer. As mentioned in previous section, there are intrinsic causes and extrinsic causes for polymer degradation. As a result, the stabilization of polymer should follow these two causes. On one hand, from the view of polymer synthesis, improvement of the polymer structure and composition should be considered; on the other hand, during polymer process, suitable additives should be added to polymers to slow down the degradation.

1.3.1 Polymer stabilization from polymerization and modification

1. Suitable polymerization method

There are many different polymerization methods used in industry, such as bulk polymerization, suspension polymerization, emulsion polymerization, coordination polymerization, etc. Different polymerization methods may result in different types of polymer structure, such as different content of unsaturated bonds and branched chains. Polymer structure is the key for polymer degradation, a careful selection of polymerization method should be considered according to different applications.

Take PVC as an example, bulk polymerization, emulsion polymerization and suspension polymerization can be used to synthesize PVC. These methods have corresponding advantages and disadvantages:

(1) Bulk polymerization. This method is very simple, low cost, and polymerization is very fast. The purity of resulted polymer is very high. The color of the resulted polymer is fine. However, due to the high concentration of monomer, the resulted PVC shows broad molecular weight distribution and higher degree of branching.

(2) Emulsion polymerization. Emulsion polymerization is very fast and can be used in continuous production. The resulted PVC shows high purity, good polymer dispersity index (PDI) and fine particle size. However, the content of unsaturated bond, branched structure and impurity is very high.

(3) Suspension polymerization. Suspension polymerization is different from the above two polymerization methods. It can produce PVC with lower content of impurity and less amount of unsaturated bonds. If the technology of polymerization is suitable, suspension polymerization can produce PVC resin with fine size, loose texture and rough surface.

As a result, suspension polymerization is broadly used in industry to prepare PVC.

2. Suitable initiator and amount

Initiator is necessary for polymerization. The polymer structures may be different in presence of different kinds and different amount of initiators. For example, titanium-containing complex is excellent initiator (catalyst) for the preparation of polyolefins. However, $TiCl_x$ may induce the decomposition of hydroperoxide and is very difficult to separate.

As shown in Table 1.5, the initiation efficiency of titanium-containing complex can be improved by modification of the catalyst structure. By decreasing the amount of titanium-containing complex, the workup and separation of the polymerization technology is simplified, the polydispersity of the resulted PP is improved, which produce polymers with higher stability.

Table 1.5 Efficiency of different kinds of Z-N catalysts

	Catalyst	PP/g	PE/g	Isotactic PE/%
1st	$TiCl_4 \cdot AlR_3$	500	300 – 400	40
2nd	$8TiCl_3 \cdot AlEt_3Cl$	3,500 – 5,000	3,000 – 4,000	89 – 91
3rd	$TiCl_3 \cdot AlR_3$-others	>300,000	1,000,000	>95

The amount of initiator is also very important for polymer stabilization. As mentioned above, initiator is a potential impurity for the resulted polymer and polymeric product. For example, in the preparation of PVC, higher amount of initiators will decrease the preparation period, increase efficiency and decrease the unsaturated bonds in PVC. However, if the concentration of initiator is more than 3 mmol/L, the thermal stability of resulted PVC will be decreased.

3. Suitable polymerization technology

There are some other factors to influence the amount of the unstable structure in polymers. The polymerization temperature, conversion, degree of polymerization as well as drying condition should be considered during the polymerization process and should be carefully selected. For example, butadiene styrene resin (SBR) can be prepared at either higher temperature (50 ℃) or lower temperature (5 ℃). For SBR prepared at higher temperature, the stability is very poor due to the higher monomer conversion, possible gelation, branched chain and poor PDI; in case of low temperature, the disproportionation is suppressed to generate polymers with low branched chain, high molecular weight and good PDI.

4. End group modification/co-polymerization/blend

Some polymers are not stable due to the reactivity on the end groups, such as POM, polymer carbonate (PC), and polysulfones. One possible solution to this problem is end group modification. For example, trace amount of water may hydrolyze POM due to the reaction on the end group. Conversion of the end group to ester bond and co-polymerization with trioxane are possible solutions (see Fig. 1.9).

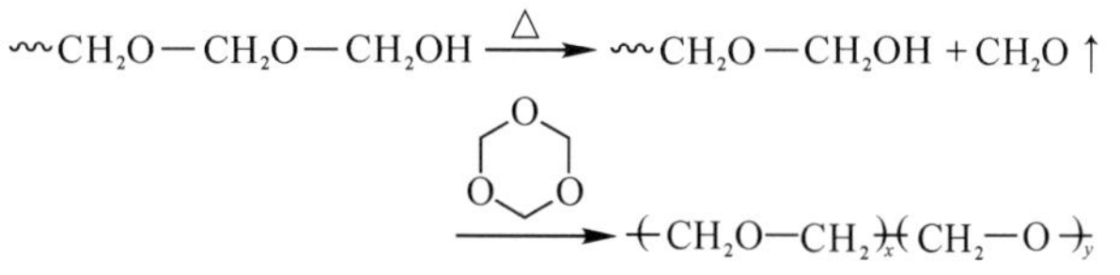

$$\sim\!\!\sim\!\!CH_2O-CH_2O-CH_2OH \xrightarrow{\Delta} \sim\!\!\sim\!\!CH_2O-CH_2OH + CH_2O \uparrow$$

$$\xrightarrow{\text{trioxane}} \pm CH_2O-CH_2\frac{}{x}CH_2-O\frac{}{y}$$

Fig.1.9 Degradation of POM and corresponding co-polymerization

The photo stability of PP fiber can be improved by co-polymerization with glycidyl methacrylate (GMA).

Blending is one of the most important methods in polymer modification. To improve the ozone stability of natural rubber, one of the possible solutions is blending with ethylene propylene diene monomer (EPDM).

5. Decrease of impurity

There are a lot of impurities in polymers which influence the stability of polymers. For example, in the preparation of PC, excess amount of monomer biphenol-A and side product NaCl may take part in the degradation of PC (see Fig. 1.10).

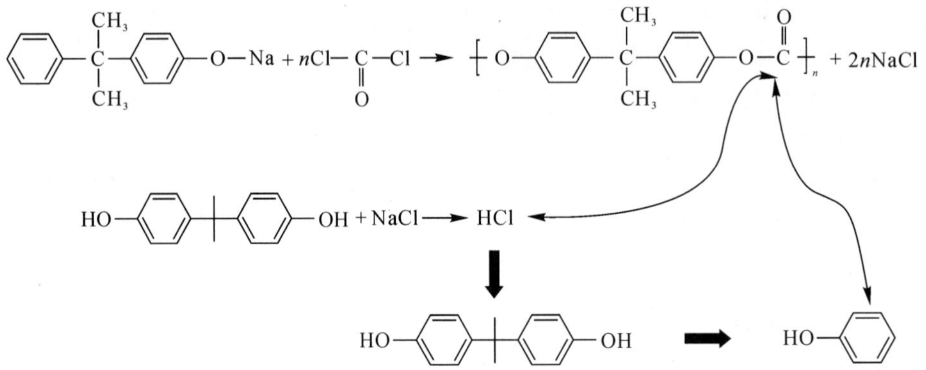

Fig. 1.10 Possible chemical reactions during the degradation of PC

Biphenol-A may react with NaCl to generate HCl, which may hydrolyze the ester bond in PC; biphenol-A may also decompose to phenol in presence of HCl, which may induce the alcoholysis of PC. These possible reactions will be accelerated at high temperature. It is very important to remove these impurities during the purification and separation process.

1.3.2 Polymer stabilization from processing and molding

Processing and molding is necessary for polymeric products. The processing of polymer is usually carried out under high temperature, strong mechanical force, and oxidation condition. If the processing of polymer is not suitable, the life time of the polymeric products will decrease a lot.

1. Pre-treatment

Pre-treatment of polymer includes drying before process and storage after preparation. As mentioned above, trace amount of water may hydrolyze the weak bonds in polymer structures. Drying is necessary before processing. Suitable drying methods should be selected according to different polymers and their processing conditions, such as drying temperature, time, etc.

2. Suitable processing instrument and temperature

Suitable processing instrument is necessary to prepare stable polymeric products. For example, homogeneous heat transport in the processing instrument should be considered. Normally, glass transition temperature plays an important role in the degradation of polymers. In contrast to small organic molecules, long-chain polymers are characterized not only by the center of mass diffusion, but also by small-scale diffusion of a few monomer units. It is generally observed that even monomolecular reactions could happen only if these motions are unfrozen. In the wider sense, the dependence of reaction efficiency on polymer morphological structures can be described in terms of the free volume concept, and of diffusion constants. These molecular characteristics are themselves dependent on the thermal transitions in polymers, the most important of which being the glass transition temperature.

The glass transition corresponds to the onset of large-scale motions of long segments of the macromolecules (10 – 20 monomer units). These liquid-like motions require more free volume, resulting in a larger volumic thermal expansion coefficient above the glass transition temperature (T_g). Below T_g, intra- and inter-molecular rearrangements are strongly hindered. This restricted mobility also limits the diffusion of oxygen, which is a controlling factor in oxidative degradation. Reactions which require a notable change in activation volume should depend much on the T_g. This is the case for the Norrish II process, in which the transition state involves a six-membered ring conformation. In the photolysis of vinyl ketone polymers and ethylene-carbon monoxide copolymers, it was observed that the quantum yield in solid film above the T_g was identical to that in solution, but decreased with temperature below T_g, to become negligible below the β-transition temperature. Reactions which involve free radicals generally require little change in activation volume and should proceed equally well below or above T_g, as is the case for the photo-fries rearrangement.

For the processing temperature of polymer, it should be 10 – 15 ℃ higher than viscous flow temperature (T_f). Therefore, long processing time of polymer at high temperature should be avoided.

3. Cooling

Besides heating, cooling is also very important for polymeric products after processing. For crystalline polymers, the cooling rate has important influence on the content of crystalline and size of crystals. In principle, upon slow cooling, the content of crystalline is high and the size of crystal is large; vise versus.

For non-crystalline polymers, fast cooling usually will generate internal stress, which may induce poor stability.

4. Post-treatment

Post-treatment of the polymeric products is also very important, which includes annealing (temperature or solvent), orientation, surface coating, and so on. Annealing can release the internal stress which is usually generated during polymer processing.

Orientation is very important for polymeric films. Oriented chains are able to increase the intermolecular interactions, and decrease the free volume and segmental mobility. Most of the orientation experiments have been performed with semicrystalline polymers (i-PP or

HDPE). In the case of mulch plastic film (PE film), both oriented draw and suitable crystalline are necessary to get transparent and stable film with higher tensile strength. Only crystalline may induce poor transparence, while only oriented draw will generated films with less stability toward heat, which may result in heat-induced shrinkage.

Surface coating is used for plastic products. For example, the surface coating of acrylonitrile butadiene styrene (ABS) resin products with chromium coating will not only improve the appearance of the products, but also improve their stability toward oxygen, water, and chemicals.

1.3.3 Additives

Addition of additives or stabilizers is one of the most important methods to improve the stability of polymers. According to different condition, the stabilizer can be classified to thermal stabilizer, photo stabilizer, antioxidant, antiozonant, etc.

1.4 Recycling of polymers and biodegradable polymers

1.4.1 Polymer recycling

Polymer recycling is very important for the protection of environment and fully use of natural resource.

As known to all, white pollution is a world-wide problem for polymers and polymeric products. For example, it is estimated that 12,800 tons of garbages is produced every day in Shanghai, while there is 130 tons of polymeric garbages. How to recycle and process polymeric garbages becomes a big problem.

On the other hand, the main source of polymer is petroleum and coal, which is non-recyclable. How to synthesize polymers very efficiently and use them in a green manner is also a big problem for a sustainable society.

1.4.2 Biodegradable polymer: definition and classification

A promising method to achieve "Green Polymer" is to use biodegradable polymers. Biodegradable polymer is defined as a material which degrades primarily by the action of microbial enzymes, which has applications in medical devices for orthopaedics, dentistry, drug release (e.g. stents), and tissue engineering. To be termed biodegradable, the rate of degradation must take place in a specified time period comparable to existing natural biodegradable materials or processes. It can also be explained as a biodegradable material which can be defined as a material that breaks down in vivo, but with no proof of its elimination from the body. For example, biodegradable polymeric systems or devices can be attacked by a biological environment so that the integrity of the material is affected and produces degradation fragments. A material shows preferential surface erosion or bulk erosion depending of its intrinsic properties (water diffusion and degradation rate) and its size. Such fragments can be carried away from their site of implantation but not necessarily from the body.

1.4.3 Principle of degradable polymers

1. Design of degradable polymers

Intrinsic causes are the key for polymer degradation. The most fundamental method for degradable polymers is design suitable structure.

As shown by Potts and his coworkers, the molecular weight of polymer is very important for polymer degradation. Polymers with higher molecular weight are very stable and cannot be decomposed by microorganism. However, if the molecular weight is lower than 500, it can be easily decomposed by microorganism. Polymers with photo sensitive groups or their blends with photosensitive polymers are very easy to photo decomposed. Polymers with —NH, —OH, —COOH and —NCO are hydrophilic, may be decomposed by microorganism.

There are lots of examples for the successful design of degradable polymers. One of them is polylactic acid (PLA), which can be decomposed by microorganism. Lactic acid can be prepared by lactobacillus, then, it can be polymerized through the following processes:

$$HO-\underset{\underset{CH_3}{|}}{\overset{\overset{COOH}{|}}{C}}-H \xrightarrow[120-160\ ^\circ C]{10-20\ \text{mmHg}①} {+}O-\underset{\underset{H}{|}}{\overset{\overset{CH_3}{|}}{C}}-\overset{\overset{O}{\|}}{C}{\xrightarrow{}}_n + nH_2O \xrightarrow[180-220\ ^\circ C]{0.1-1.0\ \text{mmHg}} \underset{O}{\overset{O}{\underset{}{\diamond}}}\longrightarrow {+}O-\underset{\underset{H}{|}}{\overset{\overset{CH_3}{|}}{C}}-\overset{\overset{O}{\|}}{C}{\xrightarrow{}}_n OH$$

PLA can be used to prepare cups and films, which can behydrolyzed in-vivo, and show broad applications in medical care.

2. Design of photodegradable polymers

The preparation of photodegradable polymers is based on the mechanism of photo degradation. The following methods can be used in practical applications:

(1) Modification of polymer. For example, addition of small amount of wax or oleic acid to PE can accelerate its photo degradation.

(2) Introduction of sensitizer or photo sensitive groups. For example, increasing the number of UV absorber groups, such as —OH and azo groups, can accelerate the degradation processes.

(3) Copolymerization with photo sensitive groups (ecolyte method). For example, the copolymerization of vinyl ketone or CO with vinyl monomers can give photodegradable PE.

---

① 1 mmHg=133.322 Pa.

3. Design of biodegradable polymers

Biodegradation of polymer is mainly due to corresponding hydrolysis induced by enzymes. There are mainly two kinds of methods for biodegradable polymers:

(1) Introduction of biodegradable monomers or related structures to polymer structure.

(2) Preparation of certain kind of biodegradable masterbatch. For example, Griffin prepared biodegradable polymers by copolymerize 80% PE and 15% starch to prepare film through blow molding.

4. Design of environmental degradable polymers

Environmental degradable polymer is referred to polymer with both photodegradable and biodegradable properties.

Questions

1. Please compare the stability of PE and PVC, and analyze the possible reasons.
2. Analyze the chemical processes during ozone cracking of natural rubber, find methods to prevent ozone cracking.
3. Can anything eat plastic? How? Why?
4. Find some polymeric products that can be degraded in your daily life. Analyze and find why and how they can be degraded.

Further reading I : polymer degradation

Most polymers will undergo significant changes over time when exposing to heat, light, or oxygen. These changes will have a dramatic effect on the service life and properties of the polymer and can only be prevented or slowed down by the addition of UV stabilizers and antioxidants. The degradation of polymers can be induced by:

(1) Heat (thermal degradation).

(2) Oxygen (oxidative- and thermal-oxidative degradation).

(3) Light (photo degradation).

(4) Weathering (generally UV/ozone degradation).

The deterioation due to oxidation and heat is greatly accelerated by stress, and exposure to other reactive compounds like ozone.

All polymers will undergo some degradation during service life. The result will be a steady decline in their (mechanical) properties caused by changes to the molecular weight and molecular weight distribution and composition of the polymer. Other possible changes include:

(1) Embrittlement (chain hardening).

(2) Softening (chain scission).

(3) Color changes.

(4) Cracking and Charring (weight loss).

Further reading II: hydrolysis of condensation polymers

Hydrolysis is the cleavage of bonds in functional groups by reaction with water. This reaction occurs mainly in polymers that take up a lot of moisture and that have water-sensitive groups in the polymer backbone. Some synthetic polymers that degrade when exposed to moisture include polyesters, polyanhydrides, polyamides, polyethers, and polycarbonates. The rate of hydrolytic degradation can vary from hours to years depending on the type of functional group, backbone structure, morphology, and pH. Polymers that readily degrade in the presence of water include polyanhydrides, aliphatic polyesters with short midblocks like polylactic acid and certain poly (amino acids) like polyglutamic acid. Assuming similar midblocks, the rate of hydrolysis decreases in the order anhydride > ester > amide > ether.

Polyanhydride $R^1\text{-CO-O-CO-}R^2 + H_2O \overset{H^+}{\rightleftharpoons} R^1\text{-COOH} + HO\text{-CO-}R^2$

Polyester $R^1\text{-COO-}R^2 + H_2O \overset{H^+}{\rightleftharpoons} R^1\text{-COOH} + HO\text{-}R^2$

Polyamide $R^1\text{-CO-NH-}R^2 + H_2O \overset{H^+}{\rightleftharpoons} R^1\text{-COOH} + H_2N\text{-}R^2$

Polyether $R^1\text{-O-}R^2 + H_2O \overset{H^+}{\rightleftharpoons} R^1\text{-OH} + HO\text{-}R^2$

Polycarbonate $R^1\text{-O-CO-O-}R^2 + H_2O \overset{H^+}{\rightleftharpoons} R^1\text{-O-COOH} + HO\text{-}R^2$

The hydrolysis of semicrystalline polymers such as esters, amides and anhydrides occurs usually in two stages. During the first stage, the degradation occurs by diffusion of water into the amorphous regions with subsequent hydrolysis. The second stage starts when moisture penetrates and degrades the crystalline regions. Thus, a bimolecular weight distribution is observed at any time during these two stages. The erosion rate usually increases over time due to decreased crystallinity and molecular weight as well as increased water solubility which turns the polymer into a beehive- or spunge-like porous body with only minor dimensional changes. Surface erosion, on the other hand, occurs at a constant erosion rate and does not affect the molecular weight but produces dimensional changes. This mechanism dominates when the polymer in question has a low water uptake and low degradation rate which is the case for many polycarbonates, polyureas, polyethers, and polyether urethanes. Surface erosion generally results in very predictable mass loss rate without compromising the mechanical and structural integrity of the polymeric materials.

There are several factors that affect the hydrolytic stability of a polymer. The most important factors are pH, temperature, hydrophobicity, morphology, degree of crystallinity and porosity. Most or all of these factors affect the water permeability and thus the bulk erosion rate. pH is one of the most important factors because an acid or a base acts as a catalyst, that is, they greatly accelerate the degradation process. The water uptake and oligomer/monomer water solubility is particularly important for polymers that degrade by bulk erosion. These factors depend on the polarity of the polymer. A lower polarity tends to decrease the reaction rate because both the water content in the polymer as well as the water permeability decrease with decreasing polarity of the polymer. Thus, the hydrolytic stability increases in the same order as the hydrophobicity. Consequently, polybutylene terephthalate (PBT) is more stable than polytrimethylene-tereph-thalate (PTT) and the latter is more stable than PET. In the case of aliphatic polyesters, the hydrolytical stability increases with the lengths of the aliphatic portion in the chain. Thus, the stability of the following aliphatic polyesters increases in the order: poly(propylene adipate) > poly(propylene succinate) > poly(propylene sebacate).

Hydrolysis reactions may be also catalyzed by certain enzymes known as *hydrolases*. These biological catalysts accelerate the reaction rate in living organisms (bacteria, fungi, etc.) without undergoing themselves any permanent change. The enzymatic hydrolysis of polymeric materials is a heterogenous process that is affected by both the physiochemical properties of the plastic [molecular weight (MW), crosslinking, chemical composition, surface area, porosity, crystallinity, etc.] and the inherent properties of the enzyme (activity, solubility, stability, 3-D conformation, etc.). Important extrinsic factors include pH, humidity, temperature, oxygen, and nutrient availability, etc.

Synthetic polymers particularly susceptible to enzyme, acid, or base catalyzed hydrolysis include aliphatic polyanhydrides like polysebacic anhydride and polyesters with short midblocks such as polylactic acid (PLA), polyglycolic acid (PGA), polycaprolactone (PCL), and polyhydroxybutyrate (PHB). These biodegradable materials find many uses in pharamceutical and medical products such as resorbable surgical sutures, resorbable orthopedic devices and controlled-release coatings for drug delivery systems.

Besides water, many other polar solvents such as alcohols, amines, and acids may cause cleavage of C—O or C—N bonds. This process is known as *solvolysis* and if the solvent is an alcohol as *alcoholysis*.

$$\sim\!\!\overset{O}{\underset{}{R^1}}\!\!\overset{O}{\underset{O}{\|}}\!\!O\!\!-\!\!R^2\!\!-\!\!O\sim + HO-R^3 \overset{H^+}{\rightleftharpoons} \sim\!\!\overset{O}{\underset{}{R^1}}\!\!\overset{O}{\underset{O}{\|}}\!\!-\!\!R^3 + HO-R^2-O\sim$$

The reaction rate depends on the type of solvolytic agent and its solubility in the polymer as well as on pH and temperature.

Further Reading III: hydrolysis of polyesters

Polyesters can undergo hydrolytic main chain scission to form water and soluble fragments. The loss of molecular weight has a dramatic effect on the service life and mechanical properties even if only one to two percent of the ester units are hydrolized. At room temperature, this reaction is rather slow, whereas at temperatures above the softening or melting point, the rate of hydrolysis increases rapidly. The effect is greatly accelerated when a catalyst like hydrochloride acid (HCl) or *p*-toluenesulfonic acid (*p*-TSA) is present. As has been shown by Szabo-Rethy et al. (1972), the critical water concentration for rapid acid catalyzed hydrolysis of polyethylene terephthalate (PET) is in the range of two percent for *p*-TSA. Thus, exposure to humidity has to be avoided during melt processing of polyesters and its copolymers whereas exposure to humidity during service life should be restricted to neutral conditions. The process of (acid catalyzed) hydrolysis of ester bonds can also be utilized to recover monomers or to tailor polymers that have to be biodegradable in moist environment (compostable plastics).

The most important commercial polyester is PET. It finds uses as fibers, films, and injection molded parts. Particularly PET bottles are of interest in regard to recycling due to the large production volumes.

The hydrolysis of polyesters is rather slow under neutral conditions (pH ≈ 7), but it is greatly accelerated when an acid or a base is present because both act as a catalyst as can be seen below.

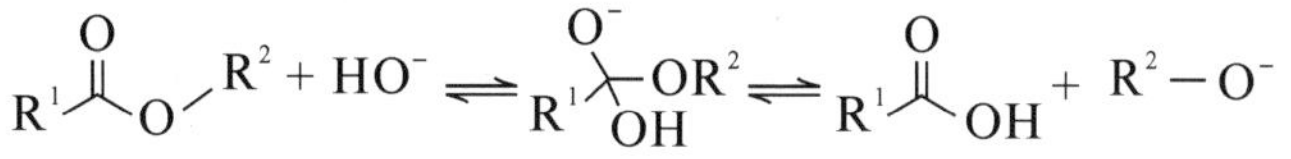

$$R^1\text{-COO-}R^2 + HO^- \rightleftharpoons R^1\text{-C}(O^-)(OH)\text{-}OR^2 \rightleftharpoons R^1\text{-COOH} + R^2\text{-}O^-$$

$$H_2O + R^2\text{-}O^- \rightleftharpoons R^2\text{-}OH + HO^-$$

Base catalyzed ester hydrolysis

$$R^1\text{-COO-}R^2 + OH_3^+ \xrightleftharpoons[-H_2O]{} R^1\text{-C}(O^+H)\text{-O-}R^2 \rightleftharpoons R^1\text{-C}^+(HO)\text{-O-}R^2 \xrightleftharpoons{H_2O}$$

$$R^1\text{-C}(HO)(H_2O^+)\text{-O-}R^2 \rightleftharpoons R^1\text{-C}(O^+H)(HO)\text{-O-}R^2 \xrightleftharpoons{-H^+} R^1\text{-COOH} + R^2\text{-OH}$$

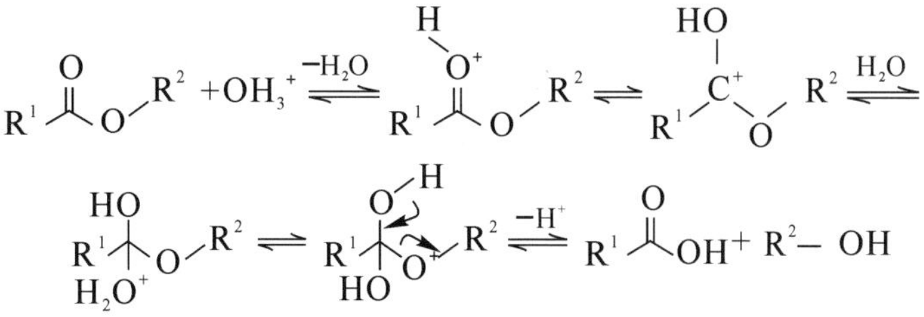

Acid catalyzed ester hydrolysis

In some cases, the chain ends are acid groups which also act as catalysts. The hydrolysis will produce additional acid groups. Thus, the reaction rate will steadily increase over time. This process is called *autocatalysis* or *auto-accelerated hydrolysis*. The reaction rate of this process is proportional to the concentration of ester groups and water as well as the concentration of acid:

$$d[COOH]/dt = k_a[H_2O] \cdot [COOR] \cdot [H_3O^+]$$

where $[H_2O]$ is the water concentration and $[H_3O^+]$ is the hydrogen ion (hydronium) concentration which depends on the overall number of acid groups. $[COOR]$ is defined as the mole equivalent of ester groups per unit volume. This simplification is only valid if we assume that all polymer chains have the same rate constants k_a regardless of size (molecular weight) of the polymer chains.

If a base is present, we also have to include a base-catalyzed hydrolysis reaction:

$$d[COOH]/dt = \{k_a[H_3O^+] + k_b[OH^-]\} \cdot [COOR]$$

In this equation, k_a and k_b are the second order rate constants of acid- and base-catalyzed hydrolysis. Since both the catalyst and the water content generally remain constant, the equation above can be simplified to:

$$d[COOH]/dt = k_h[COOR]$$
$$k_h = k_a[H_3O^+] + k_b[OH^-]$$

The product of $[H_3O^+]$ and $[OH^-]$ is constant at equilibrium. Thus,

$$K_w = [H_3O^+] \cdot [OH^-] \Rightarrow [OH^-] = K_w/[H_3O^+]$$

and

$$k_h = k_a[H_3O^+] + k_b K_w/[H_3O^+]$$

where K_w is the *ionization* or *dissociation constant* of water. This equation shows that the overall rate of hydrolysis strongly depends on pH, meaning a slight change in pH causes a significant change in the rate of reaction.

All ester reactions are equilibrium reactions, meaning the reaction will not result in completely hydrolyzed polymer unless the reaction products are continuously removed. The concentrations at equilibrium will depend on the rate constant for esterification (k_e) and depolymerization or hydrolysis (k_h). The ratio of these two constants is the equilibrium constant K:

$$K = k_e/k_h = [COO]_E[H_2O]/[COOH]_E[OH]_E = p[H_2O]/\{M_0(1-p)^2\}$$

where $[COO]_E$, $[COOH]_E$ and $[OH]_E$ are the mole equivalents of ester, acid, and alcohol groups per unit volume and p is the extend of reaction. Since the average degree of polymerization equals $X_n = 1/(1-p)$, the equation above can be rewritten as:

$$K = pX_n^2[H_2O]/M_0 = (X_n - 1)X_n[H_2O]/M_0$$

or

$$X_n \approx \{KM_0/[H_2O]\}^{1/2}$$

Thus, the equilibrium molecular weight is inversely proportional to the square root of the water content in the polymer $[H_2O]$.

Mechanical recycling can lead to some degradation due to thermal and mechanical stress which could result in lower mechanical properties and lower transparency.

Often a third term for uncatalyzed hydrolysis is included. However, this mechanism is only important at very low hydrogen ion concentration (pH≈7) because its rate constant is much smaller than those of acid- or base-catalyzed hydrolysis.

The rate of hydrolysis also depends on the stereochemical composition of the chain. The hydrogen, oxygen and carbon atoms of a group will interfere sterically with rotations of other groups which will limit the number of conformations which allow attack by hydroxyl and hydronium ions. Particularly stable are polyesters with a large number of atoms in the sixth position. This empirical rule is known as *Newman's rule of six*.

Further reading Ⅳ: biodegradable polymers utilized in food packaging

As already mentioned, packaging of foods serves to maintain their quality, freshness, and safety during distribution and storage until consumption, and for this purpose, package design and construction, as well as the selection of packaging materials and technologies, play essential roles. In this regard, it should be noticed that today's food packages often combine several materials to best exploit their main features and functions toward foods. Materials that have traditionally been used in food packaging include glass, metals, paper and paperboards, and plastics. The latter can be considered as one of the most used materials for packaging purposes. Indeed, in 2015, its consumption represented almost 40% of the total European plastic demand, which was equal to 49 Mt①. Plastics are made by polycondensation or polyaddition of monomer: there are two major categories of plastics, namely, thermosets and thermoplastics. The former are polymers that solidify or set irreversibly when heated and cannot be remolded. They are strong and durable, and for this reason, they are used primarily in automobiles and construction applications. In contrast, thermoplastics are polymers that soften on exposure to heat and return to their original condition at room temperature. These plastics can easily be shaped and molded into various forms and products, thus they are ideal for food packaging purposes. In particular, a wider variety of polymeric resins can be considered for their manufacturing: polyethylene terephthalate (PET), polyvinylchloride (PVC), polyethylene (PE), polypropylene (PP), polystyrene (PS), and polyamide (PA). Indeed, these resins have been increasingly utilized as packaging materials mainly in the light of their good mechanical performances such as tensile and tear strength, good barrier to oxygen, carbon dioxide, anhydride and aroma

---

① 1 Mt=10^6 t.

compounds, and heat sealability. Additionally, they are largely available at relatively low costs and present several desired features such as softness, lightness, and transparency, which facilitate and favor their usage and handling. However, these polymers are produced from crude oil and other fossil fuels, such as natural gas and coal, because of their chemistry lending itself to the readily accessible constituents of those fuels. In this regard, it should be highlighted that, in agreement with Colwill et al. and Ingrao et al., resource depletion is not the only one problem to be faced when dealing with synthetic plastic production, as the emission of carbon dioxide (CO_2) from fossil fuel combustion should be considered too. As a matter of fact, CO_2 is a major contributor to global warming (GW) and could have potentially devastating social, economic, and environmental consequences in the future, if not duly addressed. In addition, the massive consumption of polymeric materials is accompanied by a consistent generation of wastes that cause several environmental pollution problems.

According to Eurostat, 15.24 Mt of plastic packaging waste was generated in 2014. Plastics are mainly produced for durable scopes and, therefore, can persist undegraded for decades in the environment where they are disposed of. With regard to this point, Blanco et al. documented that their end life prediction and, hence, their persistence in the environment go well beyond the predictable one. In particular, the marine litter issue is increasingly raising great environmental concern because it is harmful to ocean ecosystems, wildlife, and humans. Besides cigarette residues, food wrappers and containers, plastic bags, beverage plastic bottles, and plastic cutlery are the most important sources of debris. In this regard, a recent study from Jambeck et al. indicated that, only in 2010, 4.8 – 12.7 Mt of plastics ended up in the oceans. In addition, as discussed in Siracusa et al., during their service, life packaging materials are often so contaminated by foodstuff and biological substances that recycling is impracticable or, most of the times, economically nonconvenient. Consequently, several tons of plastic-based materials and goods are incinerated or land filled, increasing every year the problem of municipal waste disposal. In this regard, Michaud et al. investigated the plastic waste management sector and considered a group of selected environmental impact indicators. They concluded that, where feasible, mechanical recycling is the most environmentally sustainable option as it performs best in almost all of those indicators. For contrast, incineration (with energy recovery) can be considered as the intermediary option, and landfill is confirmed as having the worst environmental performance.

Such concerns emphasize on the need for alternatives to fossil fuel-based products and processes: as a matter of fact, the continuous search for more environmentally sustainable solutions has been receiving the attention of a broad research community at the global level. In this context, the concept of eco-friendly materials has been accepted by several countries worldwide, and indeed, new advanced materials are under constant development and improvement. For instance, biopolymers with their biodegradability, eco-friendly

Chapter 1 Introduction

manufacturing processes, and vast range of types and applications can be considered as quite valid to replace the traditional, synthetic polymers on an equal function and quality basis. For this reason, the development of biopolymers has occurred largely in response to the growing concerns regarding the sustainability of conventional polymers and the environmental pollution caused by plastic packaging waste treatment. In this regard, this chapter was designed to discuss, briefly but exhaustively, about the quality- and sustainability-related issues of biopolymers, thereby exploring their feasibility of usage for food packaging purposes. Biopolymers can be considered to be mainly a solution to waste disposal problems associated with synthetic plastics and represent a loosely defined family of polymers that are designed to degrade through a series of actions performed by living organisms. Additionally, they offer a possible alternative to traditional non-biodegradable polymers where, mainly due to food contamination, recycling can neither be pursued nor is economically convenient. This is the instance case of trays made out of expanded PS and generally used for packaging of fresh foods such as meat. According to Ingrao et al., such trays act as a sponge because, for marketing and aesthetic reasons, they are conceived also to absorb the blood released by the fresh meat contained. As a consequence, they get contaminated with a variety of microorganisms that make them suitable not for being recycled but disposed of in sanitary landfills, as municipal solid wastes (MSWs). If those trays were constituted by expanded polylactic acid (PLA), or other suitable biopolymers, they could be composted along with the MSW organic fraction and other biomasses. In this way, the contamination due to the prolonged contact with the fresh meat content would no longer be a problem. Biodegradability is, in fact, one of the main reasons for the interest in bio-based polymers: it is a functional requirement and, at the same time, answers to the need and urgency of environmental sound disposal scenarios. In addition, the compostability attribute is important and required for biopolymer-based packaging products, so that they can be disposed of in compost plants and processed into compost matrixes to be utilized, in turn, as fertilizers or soil conditioners, with several environmental benefits that would occur subsequently. Moreover, in the course of the years it has been documented that production of biopolymers, especially of second generation, is less energy demanding and greenhouse gas (GHG) emitting than that of synthetic polymers.

All that stated, biopolymers could be considered valid alternative materials to those produced from fossil resources and it is quite so in nearly every function, from packaging and single use to durable products. As a matter of fact, they are gaining increasingly larger shares of the global market, thus replacing synthetic polymers in a wide range of applications, from packaging to medical, automotive, and many more.

Chapter 2 Thermal degradation and stabilization

Thermal degradation is induced by heat, which is very common in polymer synthesis and process. In the presence of oxygen, thermal degradation will undergo thermal-oxidative degradation. In case of aerospace application, thermal degradation with oxygen is also necessary to be studied.

2.1 Thermal degradation

2.1.1 Types of thermal degradation

Thermal degradation is a common phenomenon for polymers and polymeric products, a lot of researches have been studied. Polymers often undergo significant chemical changes over time when exposed to high temperatures. These changes have a dramatic effect on the service life and properties of the polymeric material. The deterioration of the physical and mechanical properties is often the result of bond breaking in the polymer backbone (chain scission) which may occur at the chain ends or at random positions in the chain. Some scientists did systematic study on thermal degradation:

(1) In 1860, C. G. Williams found that thermal degradation of natural rubber produced isoprene.

(2) In 1949, N. Grassie and H. W. Melville carried out systematic study of thermal degradation.

(3) During 1960s, W. G. Oakers, F. Chevassus, M. B. Neiman, and W. J. Bailey carried out a lot of researches on thermal degradation.

Generally, there are three different mechanisms for thermal degradation: depolymerization, unzipping reaction and chain-end scission.

In the case of depolymerization, monomers are released. This process is known as unzipping, depropagation or end-chain depolymerization. This process can be considered as the reverse process of a step-growth polymerization, monomers are removed from the polymer chains one by one. Such kind of degradation usually starts from the end of the polymer chain or weak bond of the polymer chain. At the early stage of such kind of degradation, due to the fast evaporation of monomers, the weight of polymer changes dramatically, while the molecular weight does not change too much; at the end of depolymerization degradation, since the only product is monomer, the weight of the polymer decreases close to zero, while the molecular weight change dramatically. A well-known example of end-chain depolymerization (unzipping) with high monomer yield is the decomposition of polymethyl methacrylate (PMMA). As shown in Fig. 2.1, this polymer starts to decompose at about 350 ℃ (660 °F). Random-chain fragmentation is the main initiation step in the early stages and in the later stages end-chain scission which is usually of first order. The main propagation step is unzipping to monomer which releases large amounts of methyl methacrylate (>90%). At even higher temperatures and conversion rates, decomposition proceeds by a complex series of reactions which lead to the formation of other decomposition products such as butene, methacrylic acid anhydrides.

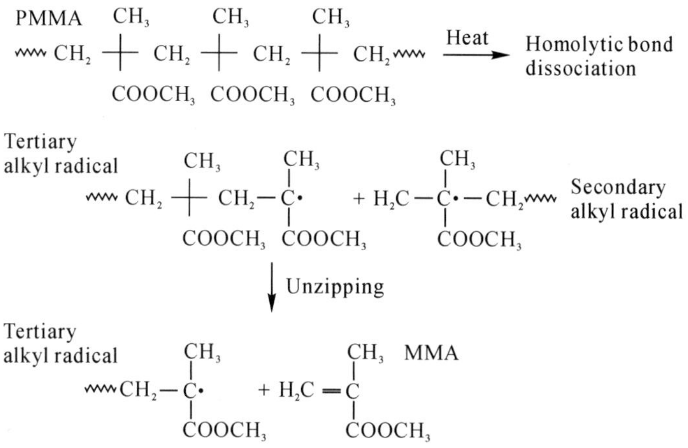

Fig. 2.1 Thermal degradation of PMMA in the way of depolymerization

1. Random chain-breaking reaction (random main chain scission)

Random main chain scission, on the other hand, leads to the formation of both monomers and oligomers (short chains with ten or fewer monomers). In such kind of degradation, the reaction site is randomly distributed on the polymer chain, while the products are low molecular weight polymers (oligomers) and monomers. Usually most polymers synthesized through polycondensation and polyaddition follow this kind of degradation. At the early stage of such kind of thermal degradation, the molecular weight decrease dramatically, the weight of polymer does not change too much; while at the end of the thermal degradation, the weight of the polymer decrease close to zero, due to the evaporation of small molecules. Thermal degradation of linear and branched polyolefins(PO) usually follows random chain-breaking reaction:

$$\text{\textasciitilde\textasciitilde\textasciitilde} CH_2-CH_2-CH_2-CH_2-CH_2 \text{\textasciitilde\textasciitilde\textasciitilde}$$
$$\downarrow$$
$$\text{\textasciitilde\textasciitilde\textasciitilde} CH_2-CH=CH_2 + H_2C=CH_2 \text{\textasciitilde\textasciitilde\textasciitilde}$$

Thermal decomposition of linear PE usually starts between 215 ℃ and 230 ℃ whereas branched PE decomposes at even lower temperatures. Usually the greater the degree of branch, the greater the rate of weight loss. Thus thermal stability decreases in the order of following sequence:

$$CH_2CH_2 > CH_2CHCH_3 > CH_2CHR > CH_2CR_2$$

The main decomposition mechanism of PE is random chain scission. Major volatile decomposition products are low molecular weight alkanes such as propane > ethane > butane, etc. which are produced by recombination reactions, and low molecular weight alkenes such as propylene > ethylene = hexene-1 > butene-1, which are produced by chain-end unzipping and molecular radical transfer followed by decomposition reactions.

In polypropylene (PP) every second atom of the polymer backbone is a tertiary carbon atom. Since these are more prone to attack than PE, PP is less thermally stable than PE. However, random chain scission and transfer does not produce much volatile compounds below 450 ℃.

Both reactions compete with cross-linking reactions, chain stripping of side groups as well as with substituent and cyclization reactions. Which of these mechanisms dominates depends on the type of polymer and temperature.

In practical application, most thermal degradation involves both of these two mechanisms. The molecular weight change and the weight change of the polymer is between these two mechanisms, as shown in Fig. 2.2.

Polymers with no or only a single (small) substituent in the repeat unit usually decompose through random chain scission rather than end-chain scission. This is the case for polyethylene, polypropylene and polymethyl acrylate (PMA). On the other hand, end-chain scission is usually the predominant decomposition mechanism in polymers with two substituents at the same carbon atom because the (large) side groups interfere with hydrogen abstraction which is known as steric hindrance. Thus disubstituted polymers like polymethyl methacrylate, poly α-methylstyrene (PMS), and polymethacrylonitrile (PMAN) usually undergo end-chain scission with high monomer yield ($\geqslant 90\%$) whereas polymers with a single large substituent are susceptible to both random chain scission and end-chain scission. However, the monomer yield usually does not exceed 50%.

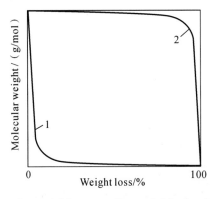

Fig. 2.2 Change of molecular weight versus the weight of polymer change for different degradation mechanisms: 1 for random chain-breaking reaction; 2 for depolymerization

Chapter 2 Thermal degradation and stabilization

Polystyrene does not undergo any appreciable weight loss below 280 ℃ though there is a reduction in molecular weight due to chain scission. The thermal decomposition process of polystyrene is rather complex including end chain scission, carbon-hydrogen bond cleavage, and free radical recombination. The main volatile products are monomer (40%–45%) with decreasing amounts of dimers, trimers, tetramers, and pentamers. The formation of monomer is mainly caused by a free radical initiated end-chain scission whereas styrene dimers, trimers, and tetramers are generated by intramolecular transfer reactions.

Poly α-styrene is less thermally stable than polystyrene. The addition of the α-methyl group provides enough steric hindrance that only monomers are generated by end-chain scission (unzipping) with a yield of about 95%.

Depolymerization involves a combination of some or all of the following reaction steps:

(1) Initiation. Depolymerization starts with the homolytic fragmentation of atoms in the chain backbone by either random chain scission or end-chain scission which produces free radicals.

$$P_n \to R_r \cdot + R_{n-r} \cdot \text{ (Random chain scission)}$$
$$P_n \to R_{n-1} \cdot + R_1 \cdot \text{ (End-chain scission)}$$

where P_n is a polymer molecule of degree of polymerization n and $R_r \cdot$ is a polymer radical of length s.

(2) Propagation. The propagation reactions of depolymerization are often called depropagation reactions. There are three different types:

$$R_n \cdot \to R_{n-r} \cdot + R_r \text{(Intramolecular H transfer)}$$
$$R_n \cdot \to R_{n-1} \cdot + M \text{ (Unzipping)}$$
$$R_s \cdot + P_n \to P_s + R_{n-r} + P_r \text{(Intermolecular H transfer)}$$

Unzipping produces monomers whereas inter- and intra-molecular hydrogen transfer usually results in fragments of larger size.

(3) Termination. The three most common termination reactions are:

$$R_n \cdot \to P_n \text{(Unimolecular termination)}$$
$$R_r \cdot + R_s \cdot \to P_{r+s} \text{(Recombination)}$$
$$R_r \cdot + R_s \cdot \to P_r + P_s \text{(Disproportionation)}$$

A first order termination reaction is theoretically not possible because it is impossible to remove a radical without adding or removing a hydrogen atom as well. However, a termination reaction can be quasi-first order if the other compound that recombines with the chain radical is so abundant that the termination reaction appears not to be affected by its concentration.

2. Small molecule elimination without breaking of the main-chain (side chain scission)

Usually, polymers with side chains (or side substituents) will be thermal decomposed through such kind of degradation mechanism. In such kind of degradation, the degradation reaction happens firstly on the side chain, the main products are the polymer backbone and small molecules (not monomer); then, the polymer backbone will decompose through either depolymerization or random chain-breaking reactions.

At the beginning of such degradation, small molecules eliminate from the side chain, while the backbone keeps its structure; at the end, the degradation will follow normal thermal degradation in either depolymerization or random chain-breaking. One of the well-known examples is the thermal degradation of polyvinyl chloride (PVC):

$$\sim CH_2-CH-CH_2-CH-CH_2-CH \sim \longrightarrow \sim CH_2-CH-CH_2-CH_2 + H_2C=CH_2 + HCl$$
$$|||||$$
$$ClClClClCl$$

$$\longrightarrow \sim CH_2-CH-C=C-C=CH\sim + HCl$$
$$||||$$
$$ClHHH$$

$$\longrightarrow \sim C=C-C=CH_2 + H_2C=C\sim + HCl$$
$$||||$$
$$HHHH$$

The thermal degradation of polyvinylalcohol (PVA) will eliminate water, the thermal degradation of poly *tert*-butyl methacrylate (PBMA) will eliminate isobutylene at the very beginning. It should be mentioned that in such kind of degradation, the reaction site of elimination of small molecules is random, can happen at any side of the polymer.

A summary of thermal degradation is as follow (see Table 2.1).

Table 2.1 Summary of different kinds of thermal degradation

	Depolymerization (Unzipping reaction)	Random chain-breaking reaction	Small molecule elimination without breaking of the main-chain
Product	Monomer	Oligomer	Small molecule (not monomer)
Early stage of degradation	Monomer evaporate quickly; molecular weight does not change too much; weight decrease dramatically	Molecular weight decrease dramatically; weight does not change too much	Side chain elimination
Certain stage of degradation	Weight decrease close to zero; molecular weight decrease dramatically	Generate lots of evaporated small molecules; weight decrease close to zero	At certain extent of the small molecules elimination, the weak site of the backbone expose, finally degradation happen on main-chain
Example	PMMA, PAMs, POM, PTFE	PE, PP, PAN, PMA	PVC, PVAc

2.1.2 Relationship between polymer structure and corresponding thermal degradation

There are relationships between the polymeric structure and the mechanisms of thermal degradation. Table 2.2 gives the summary of the thermal degradation of different kinds of polymers.

Table 2.2 Summary of different kinds of thermal degradation

Polymer	Chemical structure	Thermal degradation product	Type of degradation
PMMA	$-CH_2-C(CH_3)(COOCH_3)-$	95% monomer	Depolymerization
PAMs (poly α-methyl styrene)	$-CH_2-C(CH_3)(C_6H_5)-$	100% monomer	Depolymerization
POM	$-CH_2O-CH_2O-$	100% monomer	Depolymerization
PTFE	$-CF_2-CF_2-$	96% monomer	Depolymerization
PE	$-CH_2-CH_2-$	1% monomer, oligomer	Random chain-breaking
PP	$-CH-CH(CH_3)-$	No monomer, oligomer	Random chain-breaking
PAN (polyacrylonitrile)	$-CH_2-CH(CN)-$	No monomer, oligomer	Random chain-breaking
PMA	$-CH_2-CH(COOCH_3)-$	1% monomer, oligomer	Random chain-breaking
PVC	$-CH_2-CH(Cl)-$	HCl 95%	Small molecule elimination
PVAc (polyvinyl acetate)	$-CH_2-CH(COOCH_3)-$	HAc 95%	Small molecule elimination
PS	$-CH_2-CH(C_6H_5)-$	40% monomer Oligomer (dimer, trimer, tetramer)	Between depolymerization and random chain-breaking, more depolymerization

The following conclusions can be obtained from Table 2.2.

1. Depolymerization and polymer structure

Thermal degradation of polymers with quaternary carbon usually generates monomer with high yield, while thermal degradation of polymers that rich with hydrogen atoms generates monomer with low yield. Therefore, depolymerization follows the radical process.

For polymers with quaternary carbon radicals, it may undergo intermolecular disproportionation reaction to generate monomer.

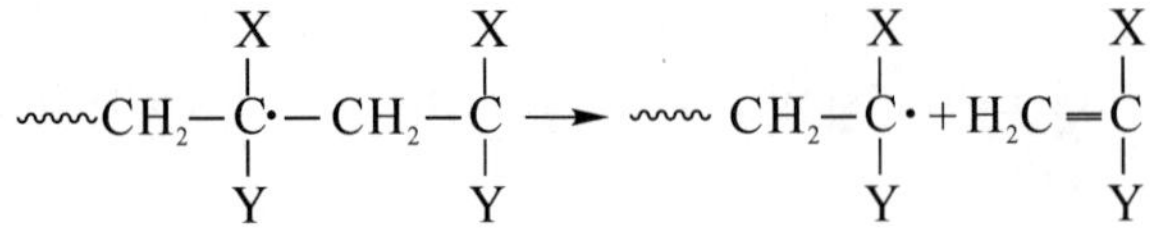

$$\sim\sim CH_2-\underset{Y}{\overset{X}{C}}\cdot-CH_2-\underset{Y}{\overset{X}{C}}- \longrightarrow \sim\sim CH_2-\underset{Y}{\overset{X}{C}}\cdot + H_2C=\underset{Y}{\overset{X}{C}}$$

If there are hydrogen atoms connected to the carbon atom, the product of the thermal degradation will undergo chain transfer reactions (see Fig. 2.3), which prevents the thermal degradation.

Intermolecular chain transfer:

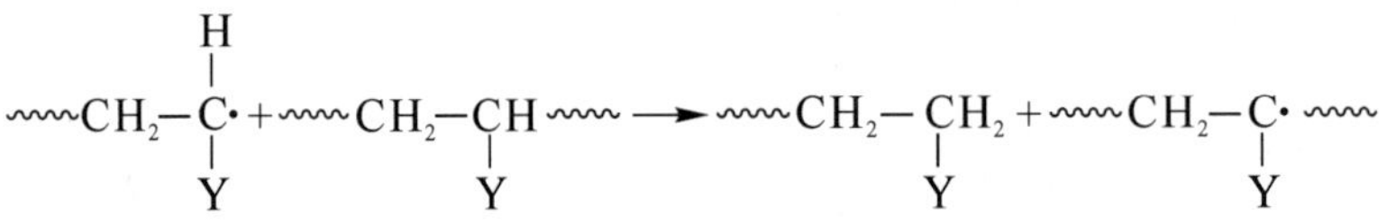

$$\sim\sim CH_2-\underset{Y}{\overset{H}{C}}\cdot + \sim\sim CH_2-\underset{Y}{CH}\sim\sim \longrightarrow \sim\sim CH_2-\underset{Y}{CH_2} + \sim\sim CH_2-\underset{Y}{C}\cdot \sim\sim$$

Intramolecular chain transfer:

$$\sim\sim CH_2-\underset{Y}{CH}-CH_2-\underset{Y}{CH}-CH_2-\underset{Y}{\overset{X}{C}}\cdot \longrightarrow \sim\sim CH_2-\underset{Y}{\overset{X}{C}}\cdot + CH_2=\underset{Y}{\overset{X}{C}}-\underset{Y}{CH_2}-\underset{Y}{CH_2}$$

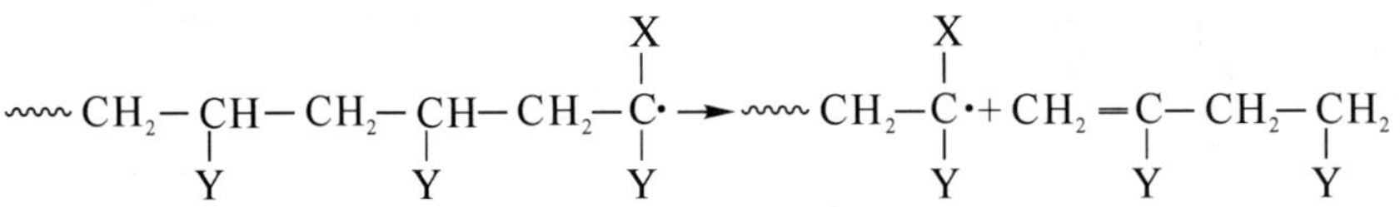

Fig.2.3 Intermolecular and intramolecular chain transfer of polymers with non-quaternary carbon radicals

2. Random chain-breaking reaction and polymer structure

Almost all of the polymers prepared from condensation and most of the polymers prepared from addition will decompose in the way of random chain-breaking reaction mechanism. The reaction site of random chain-breaking reaction is the weak site of the polymer chain. It is important to study the weak site of the polymer chain in order to well understand the thermal stability of polymers.

(1) Dissociation energy and stability. Polymer degradation usually starts from the weak site of the polymer chain. The higher the bond energy, the more difficult to break, the more stable the polymer toward thermal degradation. Table 2.3 lists the dissociation energy needed to break some chemical bonds. It is easy to find that: the bond energy in aromatic system is larger than that of the aliphatic system. Bonds with heteroatoms (N, Si, B, etc.) are also stronger than C—C bond.

Table 2.3 Typical bond energy of some chemical bonds

Bond	Energy/(kJ/mol)	Bond	Energy/(kJ/mol)	Bond	Energy/(kJ/mol)
C—C	347.5	C—C	418.7	C-C	389.4
C—O	389.4	C—O	460.5		
C—N	343.3	C—N	460.5		
C—H	410.3	C—H	431.2		

In principle, the dissociation energy can be decreased by the following effects from nearby groups: ① resonance effect. ② steric hindrance of nearby groups. ③ the valence of α-position, as shown in Table 2.4.

Table 2.4 Influence of nearby groups on the bond energy

C—C bond	C—C energy/(kJ/mol)	C—C bond	C—H energy/(kJ/mol)
CH_3—CH_3	368.4	RCH_2—H	410.3
$(CH_3)_3$—C—$(CH_3)_3$	284.7	CH_2=$CHCH_2$—H	355.9
$(C_6H_5)_3$—C—$(C_6H_5)_3$	62.8	CH_2CH_2—H	163.3

(2) Chemical structure and stability. It has been demonstrated that the unsaturation of the polymer chain and the stereoisomerism of the polymer structure have almost no influence on the thermal stability of the polymer. For example, the degradation temperature of isotactic and atactic polypropylene are the same, while isotactic and atactic polypstyrene show the same thermal stability.

The steric hindrance of the substituent may decrease the thermal stability of polymer. For example, polyethylene with different branching degree shows different thermal stability (see Fig. 2.4). The more bulky the substituent, the worse the thermal stability of the polymer.

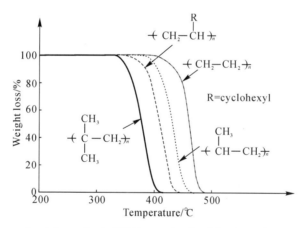

Fig.2.4 Thermogravimetric analysis(TGA) results for PE with different branching degree

(3) Crosslink and stability. It's very easy to understand that crosslink can improve the thermal stability. The higher the crosslink density, the more stable the polymer. Meanwhile, polymers with ladder, semi-ladder, network and helical structure show better thermal stability. For example, the thermal stability of polystyrene-divinyl benzene (PS-DVB) resin is dependent on the crosslink density (see Fig. 2.5). With increasing of the percentage of crosslinker (DVB) from 0% to 100%, the thermal stability of the resin

increases.

Another example is diamond. Diamond is crosslink non-aromatic crystal. In inert atmosphere, it is stable up to 1,300 ℃. By increasing temperature to 1,500 ℃, it will form aromatic graphite. Graphite is very stable until 2,400 ℃. Such kind of stability is attributed to the crosslink and aromatic structure of graphite.

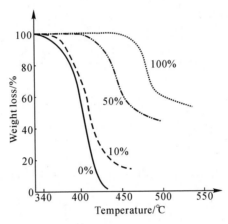

Fig.2.5 TGA results for styrene-DVB resin with different crosslink density

(4) Crystallinity and stability. According to Mark Triangle Principle (see Fig. 2. 6), the increasing of the crystallinity of the polymer can increase the thermal stability of the polymer. For practical application, it should be balanced for the crystallinity, crosslink and chain rigidity of the polymer to get better thermal stability of polymeric products.

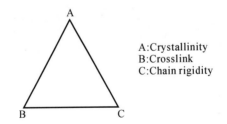

Fig.2.6 Mark triangle principle for polymers

2.2 Stabilization of thermal degradation

According to the above analysis for thermal degradation, there are several methods for the stabilization of thermal degradation.

2.2.1 Addition of thermal stabilizer

A direct method to improve the thermal stability of polymer is addition of corresponding

Chapter 2 Thermal degradation and stabilization

thermal stabilizer, which is usually added during the processing and molding period. Generally, there are two kinds of thermal stabilizers according to different mechanism: ① react with the active bonds in polymer to give stable bond. For example, the stability of POM can be improved by addition of suitable agents to form ester bond or other bond, which is stable than the terminal hydroxyl group. ② stop the chain reaction of thermal decomposition. For example, in the depolymerization of PMMA, addition of radical quencher can stop the radical process. In **Chapter 7**, we will introduce different kinds of thermal stabilizers used for certain kind of polymer system.

2.2.2 Optimize the polymer structure

For application in aerospace, only addition of thermal stabilizer cannot prepare polymers with enough stability, one possible solution is to optimize the structure of the polymer. For example, Nomex (poly-m-phenylene terephthamide, HT-1) is stable up to 230 ℃. However, due to the hydrogen bond in such kind of polymeric system, these polymers are very difficult to process. For example, to prepare processable polyimides (PIs) products from insoluble PIs, the soluble polyamic acid (PAA) is firstly synthesized from the condensation between pyromellitic dianhydride (PMDA) and diamine, followed by thermal dehydration (see Fig. 2.7).

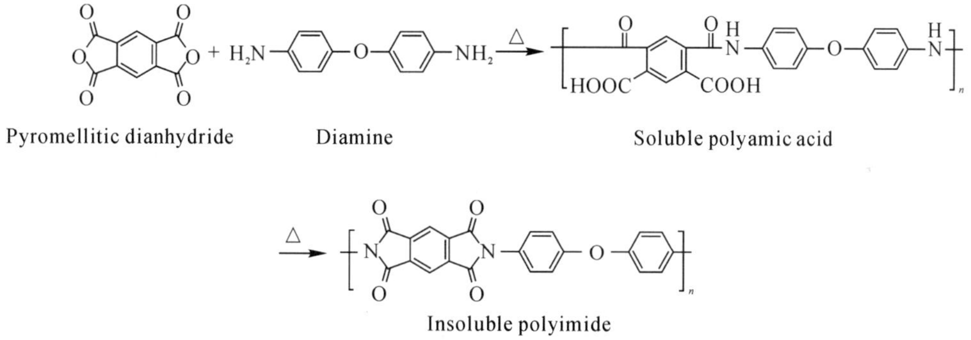

Fig.2.7 Synthetic approach of insoluble polyimides

Formation of ladder or helical structured polymer is another strategy for high thermal stability polymers. For example, thermal treatment of polyacrylonitrile (PAN) can form aromatic structures upon 500 ℃, increasing temperature to 800 ℃, the aromatic structure will undergo dehydrogenation to form ladder-like structure, which is more stable:

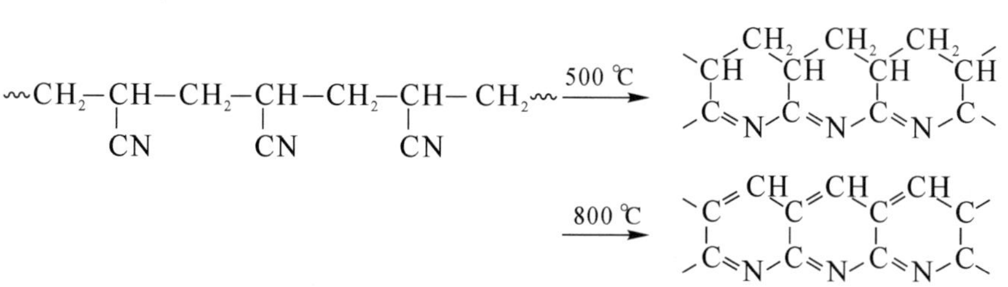

2.3 Thermal-oxidative degradation

2.3.1 Mechanism of thermal-oxidative degradation

During the preparation, storage and application, polymers and polymeric products will definitely contact with oxygen. At certain temperature, the presence of oxygen will accelerate the degradation process. The thermal degradation involved with oxygen is called thermal-oxidative degradation, which is more common than thermal degradation. HDPE is stable up to 290 ℃ in anaerobic condition; while it will decompose up to 100 ℃ in the air. J. L. Bolland and G. Gee studied the thermal-oxidative degradation of polyolefin (see Fig. 2.8). At the beginning of the reaction A-B, it's a linear relationship; the slope increases during B-C; reach saturated during C-D. The absorption of oxygen starts from A for pure polymer; if there is some auxiliary in the polymer, the reaction will have an induction period (O-A). The process of B-C is an automatic oxidation process. After the absorption of oxygen, the thermal-oxidative degradation will happen. Most of the thermal-oxidative degradations of polymer usually shows "S" shape characteristic.

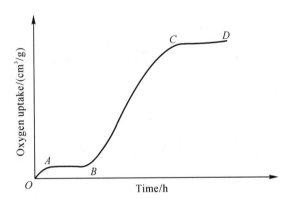

Fig.2.8 Oxygen absorption curve of polyolefin

The automatic oxidation process is the main characteristics of thermal-oxidative degradation, also the core of thermal-oxidative degradation. The general mechanism of thermal-oxidative degradation of polymers is shown in Fig. 2.9.

There are several elementary reactions for thermal-oxidative degradation.

(1) Initiation. Oxidative degradation is usually initiated when polymer chains form radicals, either by hydrogen abstraction or by homolytic scission of a carbon-carbon bond. This can occur during manufacture, processing or during service when exposing the polymer to light or heat.

$$R-H \rightarrow R\cdot + H\cdot$$

(2) Propagation. The propagation of thermal degradation involves a number of reactions.

Chapter 2 Thermal degradation and stabilization

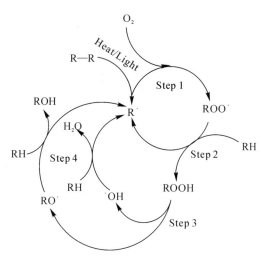

Fig. 2.9 General mechanism of thermal degradation

The first step is the reaction of a free radical (R·) with an oxygen molecule (O_2) to form a peroxy radical (ROO·), which then abstracts a hydrogen atom from another polymer chain to form a hydroperoxide (ROOH). The hydroperoxide splits then into two new free radicals, (RO·) + (·OH), which abstract labile hydrogens from other polymer chains. Since each initiating radical can produce two new free radicals, the process can accelerate depending on how easy it is to remove hydrogen from other polymer chains and how quickly free radicals undergo termination via recombination and disproportionation.

$$R· + O_2 \rightarrow ROO·$$
$$ROO· + RH \rightarrow R· + ROOH$$
$$ROOH \rightarrow RO· + ·OH$$
$$RO· + RH \rightarrow R· + ROH$$
$$·OH + RH \rightarrow R· + H_2O$$

It should be mentioned that the accumulation of ROOH will generate more radicals. The RO· radical can also undergo disproportionation to form a carbonyl compound and a new radical.

$$\sim\sim CH_2-\underset{\underset{C_6H_5}{|}}{\overset{\overset{CH_3}{|}}{C}}-CH_2-\overset{\overset{CH_3}{|}}{CH}\sim\sim \longrightarrow \sim\sim CH_2-\underset{\underset{O·}{\|}}{\overset{\overset{CH_3}{|}}{C}}-CH_2 + \sim\sim \overset{\overset{CH_3}{|}}{CH}-\overset{·}{C}H_2$$

(3) Termination. Termination of thermal degradation is achieved by recombination of two radicals or by disproportionation/hydrogen abstraction. These reactions always occur but can be accelerated by addition of stabilizers.

Recombination of two chain radicals results in an increase of the molecular weight and crosslinking density:

$$R· + R· \rightarrow R-R$$

$$2ROO\cdot \to ROOR + O_2$$
$$R\cdot + ROO\cdot \to ROOR$$
$$R\cdot + RO\cdot \to ROR$$
$$HO\cdot + ROO\cdot \to ROH + O_2$$

The result of these reactions is enbrittlement and cracking of the polymer. Termination by chain scission, on the other hand, results in the decrease of the molecular weight leading to softening of the polymer and reduction of the mechanical properties.

$$R_n\cdot + R_m\cdot \to R_n-CH=CH_2 + R_m$$
$$2RCOO\cdot \to RC=O + ROH + O_2$$

Which of these termination steps is predominant will depend on the type of polymer and on the conditions. For example, polyolefins with short alkyl side groups like polypropylene and polybutylene, and unsaturated polymers like natural rubber (polyisoprene) undergo predominantly chain scission, whereas polyethylene, and rubbers with somewhat less active double bonds like polybutadiene and polychloropene suffer from embrittlement due to crosslinking during aging.

2.3.2 Equation of thermal-oxidative degradation

The key reaction of thermal-oxidative degradation is the accumulation and decomposition of ROOH, this process is balanced. The concentration of ROOH will reach highest when the accumulation rate is equal to the decomposition rate. Then the reaction will slow down and reach a stable state. In 1963, Shelton found that the decomposition of ROOH is a first-order and second-order reaction. In solid state, the ROOH form randomly on the polymeric chain, and follow first-order reaction in the very beginning of automatic oxidation; after the accumulation of ROOH, bimolecular decomposition becomes dominant. To simplify the reaction, we only consider the second-order reaction for the automatic oxidation process.

(1) Initiation.
$$2ROOH \to RO\cdot + ROO\cdot + H_2O$$

(2) Propagation.
$$R\cdot + O_2 \to ROO\cdot$$
$$ROO\cdot + RH \to R\cdot + ROOH$$

(3) Termination.
$$R\cdot + R\cdot \to R-R$$
$$2ROO\cdot \to ROOR + O_2$$
$$R\cdot + ROO\cdot \to ROOR$$

If the $[O_2] = 21\%$ (same in air) or higher, the reaction between $R\cdot$ and O_2 is very fast, then $[R\cdot] \ll [ROO\cdot]$. Therefore, the combination termination of $ROO\cdot$ is dominant. At equilibrium state ($v_i = v_t$), the concentration of $ROO\cdot$ can be calculated:

$$v_i = v_t = 2k_t[ROO\cdot]^2$$
$$[ROO\cdot] = (v_i/2k_t)^{1/2}$$

The rate of oxidation v_{ox} only dependent on the first reaction of the propagation:
$$ROO\cdot + RH \longrightarrow ROOH + R\cdot$$
$$R\cdot + O_2 \longrightarrow ROO\cdot$$
Thus
$$v_{ox} = -d[O_2]/dt = d[ROOH]/dt = k_p[ROO\cdot][RH] = k_p(v_i/2k_t)^{1/2}[RH]$$

We can draw the conclusion that: at higher $[O_2]$, v_{ox} is not dependent on $[O_2]$, is dependent on $[RH]$.

If the $[O_2]$ is very low, the oxidation rate is not dependent on the reaction between $R\cdot$ and polymer, then $[R\cdot] \gg [ROO\cdot]$, the oxidation rate is dependent on the reaction rate between $R\cdot$ and O_2. Therefore, the bimolecular termination of $R\cdot$ is dominant. At equilibrium state ($v_i = v_t$), oxidation rate can be calculated:
$$v_i = v_t = 2k'_t[R\cdot]^2$$
$$[R\cdot] = (v_i/2k'_t)^{1/2}$$
$$v_{ox} = -d[O_2]/dt = k'_p[R\cdot][O_2] = k'_p(v_i/2k'_t)^{1/2}[O_2]$$

We can draw the conclusion that: at lower $[O_2]$, v_{ox} is dependent on $[O_2]$, not dependent on $[RH]$.

2.3.3 Factors impact thermal-oxidative degradation

The thermal-oxidative stability is different for different kinds of polymers, even for the same polymer under different kinds of conditions, their thermal-oxidative properties are different. In generally, there are several factors that impact the thermal-oxidative degradation, such as degree of saturation of polymer, degree of branch of polymer, substituent and crosslink, crystallinity, metal ions, etc.

1. Degree of saturation of polymer

It is found that the saturated polymer is more stable than unsaturated diene-containing rubber under thermal-oxidation condition. Such a difference in thermal-oxidative stability is related to the diene structure, which can impact the adjacent C—H bond energy and activation energy in the radical chain propagation. For example, the dissociation energy of C—H in phenyl (C_6H_5—H) and vinyl ($CH_2=CH$—H) is 435 kJ/mol, while the corresponding energy for ethylene propylene group ($CH_2=CHCH_2$—H) and benzyl ($C_6H_5CH_2$—H) is only 356 kJ/mol. The decreasing of the dissociation energy is due to the improved stability of radical, which is stabilized by the conjugation of the π-system.

Activation energy (ΔE_p) is needed to overcome for polymerization reaction in the radical way. The less the activation energy, the easier the reaction. It is found that the unsaturated structure can activate the proton at the α-position. Therefore, other unsaturated structures such as carbonyl and nitrile can activate the proton at the α-position and stabilize the radical, then decrease the dissociation energy.

2. Degree of branch of polymer

As mentioned in previous section, the thermal-oxidative stability of linear polymer is better than the branched one. The oxidation activity of C—H bond in polymer system is the same as that in organic chemistry. The oxidation activity of tertiary carbon is higher than the primary and secondary carbon. Similarly, the introduction of alkyl chain to the polymer backbone will generate branched polymer. As a result of the improved oxidation activity, the thermal-oxidative stability becomes worse.

$$\sim\sim CH_2-CH_2-CH_2-CH_2-CH_3 \xrightarrow{-C_nH_{2n+1}} \sim\sim CH_2-\underset{\underset{C_nH_{2n+1}}{|}}{CH}-CH_2-\underset{\underset{C_nH_{2n+1}}{|}}{CH}-CH_3$$

3. Substituent and crosslink

The substituents can also impact the thermal-oxidative stability of polymer. The influence is very complicate, it should be discussed by different groups. For example, the reactivity of the benzyl C—H is similar even higher than the tertiary one, however, the thermal-oxidative stability of polystyrene (benzyl C—H) is higher than polypropylene (tertiary C—H), which is due to the steric hindrance protection of phenyl ring to the polymer backbone.

Bolland carried out a systematic study on the substituent effect on hydrogen abstraction activity of propylene ($CH_3-CH=CH_2$) at 45 ℃, and got the following rules:

(1) If there are some substituents on $-CH_3$ or $-CH_2-$, one H is substituted by R, the possibility of oxidation increase by 3.3^n, n is the number of substituent.

(3) If one H in $-CH_3$ is substituted by phenyl, the reactivity of oxidation increase by 23 times.

(3) If one H in $-CH_3$ is substituted by vinyl, the reactivity of oxidation increases by 107 times.

Crosslinked structure can definitely increase the thermal-oxidative stability of polymer. In the crosslinked polymeric structures, the crosslink bond can avoid the degradation due to some oxidation-induced bond cleavage. Polymers with network, ladder-like and semi ladder-like structures show improved thermal-oxidative stability.

4. Crystallinity

Crystal is ordered structure, the molecules are packed in a long period. With regarding to thermal-oxidative degradation, the diffusion and penetration of oxygen into the polymeric structure is a key issue. In the crystalline domain, the diffusion of O_2 is very slow due to the closed packing, which results in a better thermal-oxidative stability. For example, linear polyethylene (PE) shows higher crystallinity than branched PE, which also results in a better thermal-oxidative stability of linear PE. If heat both polymers above their melting point, the rate of oxidation of these two polymers are very close to each other.

In practical application, polymers with antioxidants will exclude the antioxidants to the amorphous region when they crystalize from the melting state. This is very useful for anti-oxidation.

5. Metal ions

The presence of metal ions cannot be avoided, especially the presence of transition metal

Chapter 2 Thermal degradation and stabilization

ions, will increase the rate of automatic oxidation. Cu, Fe, Co, Ni and Mn are common metal ions for polymers, which are very important in the polymers used in electric wires and electric cables. Metal ions can be introduced during the polymerization process from the containers and pipes, polymer processing and molding from the instruments, as well as the transfer of metal ions during application, such as transfer of Cu in the electric cables.

G. Scott found that metal ions may aid the oxidation process through one electron oxidation process. As shown in Fig. 2.10, metal ions can form unstable complex with ROOH, which will undergo electron transfer to form radical and catalyze the automatic oxidation process. The metal ions can be used as either reductant or oxidant during this process.

As reductant:
$$ROOH + M^{n+} \longleftrightarrow [M-ROOH]^{n+} \rightarrow RO\cdot + OH^- + M^{(n+1)+}$$

As oxidant:
$$ROOH + M^{(n+1)+} \longleftrightarrow [M-ROOH]^{(n+1)+} \rightarrow ROO\cdot + H^+ + M^{n+}$$

Fig.2.10 Possible reaction of metal ions during the automatic oxidation of polymer

The possibility of both reactions shown in Fig. 2.10 is dependent on the relative activity of the reductant or oxidant role of the metal ions. For example, if the metal ions have more than one stable valence, both reactions are possible, for example Co^{2+} and Co^{3+}:

$$2ROOH \xrightarrow{Co^{2+}/Co^{3+}} RO\cdot + ROO\cdot + H_2O$$

It is easy to find that trace amounts of metal ions will convert large amounts of ROOH to radicals, and further increase the reaction rate. Minoru Imoto introduced another mechanism (see Fig. 2.11). In the presence of Fe^{2+}, it will donate one electron to ROOH and convert Fe^{2+} to Fe^{3+}, and then increase the decomposition rate of ROOH.

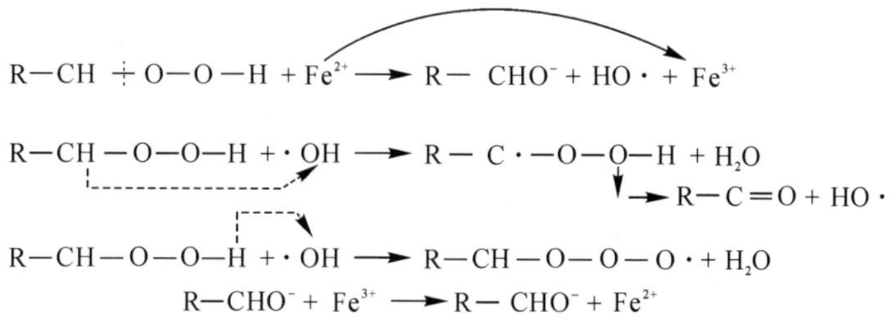

$$R-CH \dotplus O-O-H + Fe^{2+} \longrightarrow R-CHO^- + HO\cdot + Fe^{3+}$$
$$R-CH-O-O-H + \cdot OH \longrightarrow R-\overset{\cdot}{C}-O-O-H + H_2O$$
$$\longrightarrow R-C=O + HO\cdot$$
$$R-CH-O-O-H + \cdot OH \longrightarrow R-CH-O-O\cdot + H_2O$$
$$R-CHO^- + Fe^{3+} \longrightarrow R-CHO^- + Fe^{2+}$$

Fig.2.11 Possible reaction of Fe^{2+} with ROOH

In conclusion, metal ions will accelerate the decomposition of ROOH to form radical and increase the initiation rate. The activity of metal ions as reductant or oxidant is related to their coordination ability and redox potential of their complexes with ROOH. Thus, to avoid the influence of metal ions, there are generally two strategies: ① stabilized metal ions to their highest coordination number. ② change metal ions to insoluble species.

2.4　Stabilization of thermal-oxidative degradation

According to the above analysis, rational design the structure of polymers is a powerful tool to improve the thermal-oxidative stability of polymers. However, there is some limit for commercial polymers. Design of new structures will not only increase the cost, but also lose some special properties of these polymers. Therefore, the addition of antioxidant to polymers and polymeric products is the general and useful method. The selection and use of antioxidants should be very careful according to their structures, properties, as well as their storages and application conditions.

According to their mechanisms, the antioxidants can be classified to two different categories: primary and auxiliary.

Primary antioxidant is also called radical scavenger. It can react with the radicals (ROO·, RO·, ·OH, etc.) during auto-oxidation and stop the chain propagation reaction. It's obvious that the addition of primary antioxidant can change the automatic oxidation process.

Auxiliary antioxidant is also called preventive antioxidant. It can decompose the ROOH species formed during auto-oxidation process. It's obvious that the addition of auxiliary antioxidant can only slow down the oxidation rate, cannot change the automatic oxidation in mechanism.

In principle, the earlier addition of antioxidant, the better the performance. However, it's not very common to add antioxidant during the synthesis of polymer due to its possible influence the polymerization reaction. For example, in the suspension polymerization to prepare PVC, antioxidant is added with the monomer. While for acrylonitrile butadiene styrene plastic (ABS) and polyolefin, the addition of antioxidant is usually in the post-treatment process, which can avoid their decomposition during the separation and drying process.

Most antioxidants are added during the batching process, such as mix (or blend) polymers, pelleting process. It will not only prevent the oxidation of polymer during such process, but also prevent the thermal-oxidative degradation during polymer processing and molding as well as the storage process.

Antioxidant can also be added during the polymer processing, for example, mix antioxidant with other auxiliary to prepare masterbatches.

2.4.1　Primary Antioxidant

The primary antioxidant can be classified into four different types: ① proton donor. ② electron donor. ③ radical scavenger. ④ benzofuranone.

1. Proton donor

Secondary aromatic amine and frustrated phenol belong to the proton donor type primary antioxidant. The reactive —NH and —OH can compete with polymer to react with

Chapter 2 Thermal degradation and stabilization

ROO·, and form ROOH and stable radical $Ar_2N·$ or ArO· through hydrogen transfer (see Fig. 2.12), which will stop the propagation process. The formed $Ar_2N·$ and ArO· can also react with ROO· and stop the possible reaction.

Secondary aromatic amine:

$$Ar_2N-H + ROO· \longrightarrow ROOH + Ar_2N·$$
$$Ar_2N· + ROO· \longrightarrow Ar_2NOOR$$

Frustrated phenol:

$$ArOH + ROO· \longrightarrow ROOH + ArO·$$
$$ArO· + ROO· \longrightarrow ROOArO$$

Fig.2.12 Possible reaction of proton donor type antioxidants

The most famous antioxidant is 2,6-*di-tert*-butyl-4-methylphenol (butylated hydroxytoluene, BHT) (see Fig. 2.13). It can react with ROO·, the formed phenyl radical can be stabilized through resonance. BHT can protect the polymers from heat, oxidation as well as Cu^{2+}, and widely used in PVC, PE, PP, PS, ABS, PET, etc, and may be used for food package. The disadvantage of BHT is the yellowing of polymer and its volatile behavior.

N,N'-bis(2-naphthyl)-p-phenylenediamine is a widely used secondary aromatic amine type antioxidant. It displays good antioxidant property as well as good thermal stabilization and inhibition of the effect of metal ions. It can be used in the polymeric product from PE, PP, PS, ABS, POM, etc.

Reaction with radical:

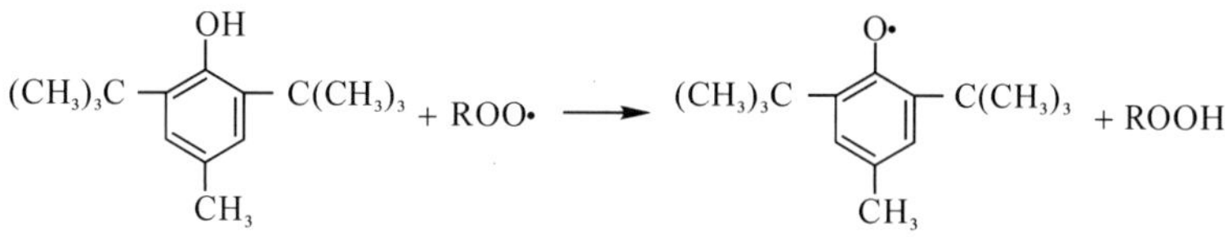

Resonance to stabilize:

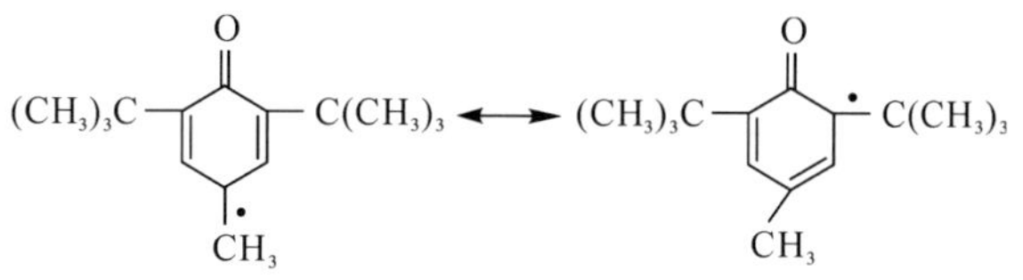

Reaction with radical:

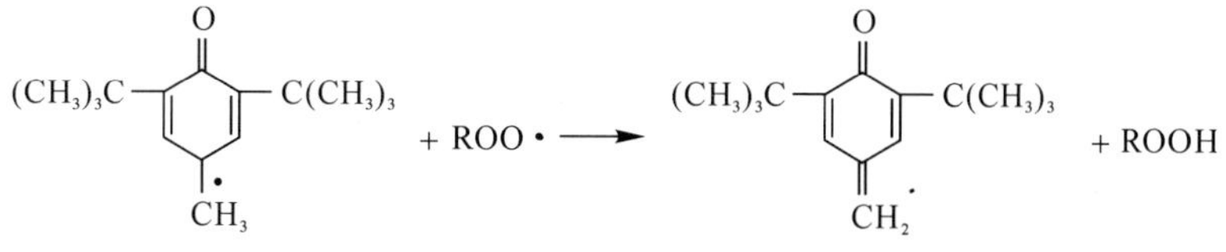

Fig.2.13 Possible reaction of BHT during thermal-oxidative process

2. Electron donor

Tertiary amine without reactive N—H belongs to the electron donor type antioxidant. The mechanism of such kind of antioxidant is not very clear. Generally, this kind of antioxidant can donate electron, and form Wurster's radical cation. The application of such kind of antioxidant is not very common.

$$ArNR_2 + ROO \cdot \rightarrow ROO \cdot - ArN \cdot R_2^+$$

3. Radical scavenger

Since the thermal-oxidative process is a radical process, any compound that can react with radicals can stop the automatic oxidation reaction and stop the thermal-oxidative degradation. Theoretically, those compounds that can capture radicals and the corresponding products will not initiate any radical chain reaction is called radical scavenger. For example, hydroquinone is used as inhibitor during the storage and transportation of monomers due to its radical scavenger property. The quinone can react with radicals in the following ways (see Fig. 2.14).

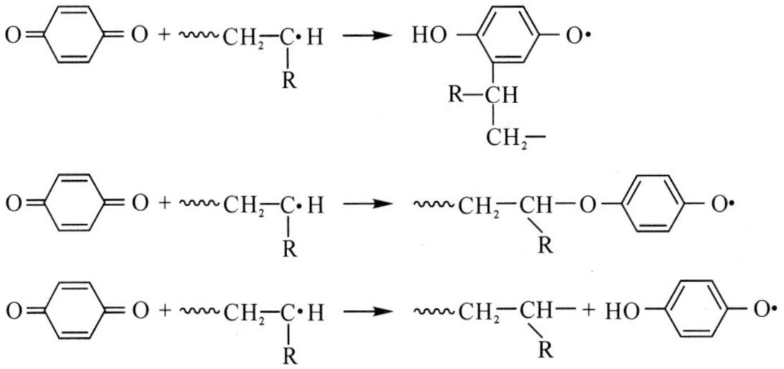

Fig. 2.14 Possible reaction of quinone with radicals

The formed new radicals in this process cannot initiate new chain reaction, can only undergo coupling termination or disproportionation termination (see Fig. 2.15).

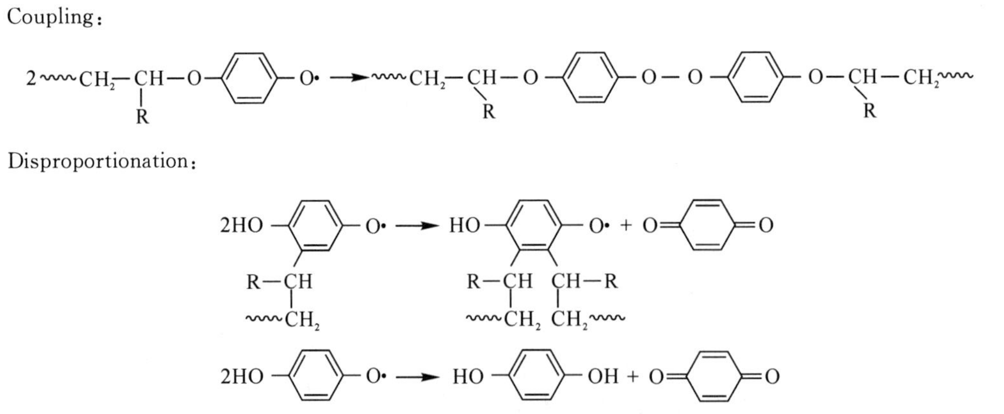

Coupling:

Disproportionation:

Fig. 2.15 Termination reaction of the new radical formed by reaction of quinone with radicals

Chapter 2 Thermal degradation and stabilization

Carbon black and polynuclear aromatic hydrocarbons can also be used as radical scavenger. For example, carbon black is used as stabilizer in rubber products. The structure of polynuclear aromatic hydrocarbons is very complicate (Fig. 2.16), there are some proton, phenol, quinone and carbonyl groups on the edge of the planar structure of polynuclear aromatic hydrocarbons. The phenol can be used as proton donor to stop the oxidation reaction, while the quinone and polynuclear aromatic structure can capture the radical and stop the chain reaction.

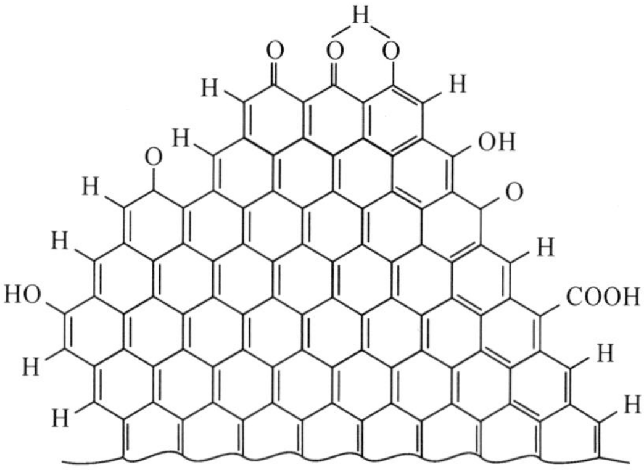

Fig. 2.16 Chemical structure of polynuclear aromatic hydrocarbons

4. Benzofuranone

Benzofuranone is a new kind of antioxidant. Actually quinone type antioxidants can only be used as inhibitor not real antioxidant in practical applications. The other two kinds of antioxidant can only stop the ROO·, cannot capture the R·. However, in the condition of very low concentration of O_2, the removal of R· is very important.

Benzofuranone type antioxidant can capture two macromolecular radicals (see Fig. 2.17). In the first step, it is used as a proton donor to capture the first macromolecular radical; in the second step, it will react with a second macromolecular radical.

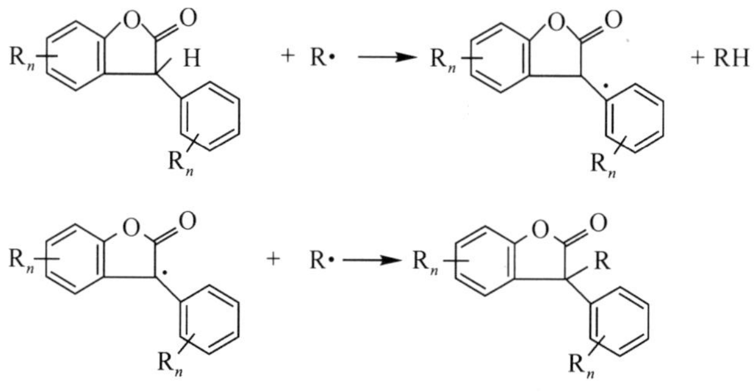

Fig. 2.17 Possible reactions of benzofuranone with radicals

We can find that the stability of radical formed in the first step in necessary to enable the second reaction. The steric hindrance, electronic resonance and the donating and withdrawing property of the substituent is related with the stability of the radical. Benzofuranone matches all of these structure characteristics. The cost performance of benzofuranone can be enhanced by combination with other type of antioxidants.

2.4.2 Auxiliary antioxidant

Auxiliary antioxidant can be used to decompose the ROOH species without the formation of radicals. There are two kinds of auxiliary antioxidants: phosphite and organosulfur compound.

1. Phosphite $[(R_1O)_3P]$

In the structure of phosphite, R_1 can be alkyl, aromatic and other types of substituents. Phosphite can reduce the ROOH to corresponding alcohol, while itself been oxidized to phosphate. Meanwhile, it can also react with ROO· and RO· radicals (see Fig. 2.18). In practical applications, phosphite is usually used together with phenol-type primary antioxidants.

$$\text{Reduce ROOH to ROH: } P(R_1O)_3 + ROOH \rightarrow ROH + O=P(OR_1)_3$$
$$\text{React with ROO· and RO·: } P(R_1O)_3 + ROO· \rightarrow RO· + O=P(OR_1)_3$$
$$P(R_1O)_3 + RO· \rightarrow R· + O=P(OR_1)_3$$

Fig. 2.18 Possible reactions of phosphite during thermal-oxidation process

2. Organosulfur compound (R—S—R)

Thioether can convert two ROOH to ROH through stepwise oxidation (see Fig. 2.19).

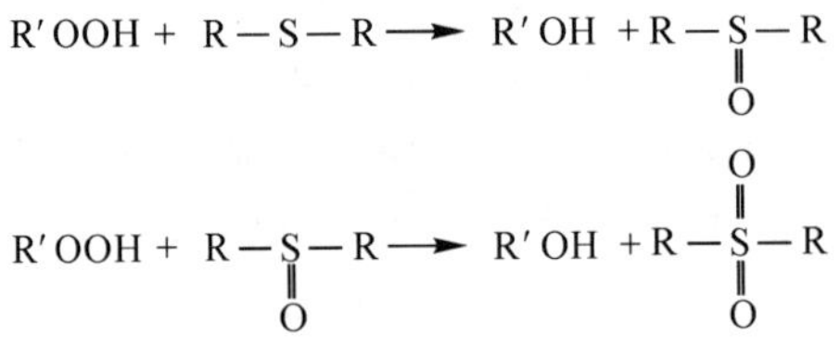

$$R'OOH + R-S-R \rightarrow R'OH + R-\underset{\underset{O}{\|}}{S}-R$$

$$R'OOH + R-\underset{\underset{O}{\|}}{S}-R \rightarrow R'OH + R-\underset{\underset{O}{\|}}{\overset{\overset{O}{\|}}{S}}-R$$

Fig. 2.19 Possible reactions of thioether with ROOH

In fact, there are some other possible reactions (see Fig. 2.20), such as the formation of sulfenic acid, thiosulfinates and hyposulphuric acid.

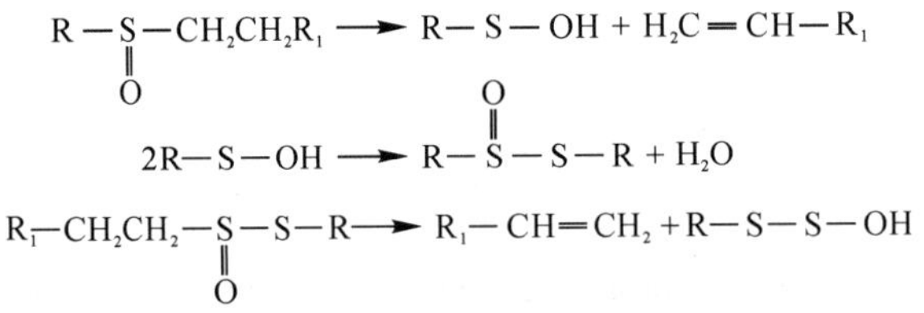

$$R-\underset{\underset{O}{\|}}{S}-CH_2CH_2R_1 \rightarrow R-S-OH + H_2C=CH-R_1$$

$$2R-S-OH \rightarrow R-\underset{\underset{O}{\|}}{S}-S-R + H_2O$$

$$R_1-CH_2CH_2-\underset{\underset{O}{\|}}{S}-S-R \rightarrow R_1-CH=CH_2 + R-S-S-OH$$

Fig. 2.20 Other possible reactions of thioether during thermal-oxidation degradation

Chapter 2 Thermal degradation and stabilization

It is found by G. Scott that sulfur-containing organic acids are antioxidant with catalytic property towards ROOH, while their corresponding oxidation products (SO_2, SO_3) are effective decomposition agents for ROOH.

2.4.3 Combination of antioxidant

In practical applications, the primary and auxiliary antioxidants can be used separately or together. There are three different kinds of situations for the combination of different kinds of antioxidants: synergetic, additive, and antagonistic.

1. Synergetic effect （1＋1＞2）

Synergetic effect means that the effect of combination of two or more antioxidants is better than separately use. It also includes homogeneous and heterogeneous synergetic effect according to antioxidation mechanism. For example, the combination of frustrated phenol and amine is homogeneous synergetic effect, because both antioxidants work as proton donor to ROO·. In the case of combination use of different kinds of phenol or different kinds of amines or one phenol with one amine, the synergetic effect arise from the hydrogen transfer. Highly effective antioxidant will first react with ROO· to stop the chain reaction and generate an antioxidant radical; the low effective antioxidant will donate proton to the radical to form new antioxidant molecule and regenerate the highly effective antioxidant, meanwhile the formed low effective antioxidant radical can also react with ROO· to stop its reactivity (see Fig. 2.21).

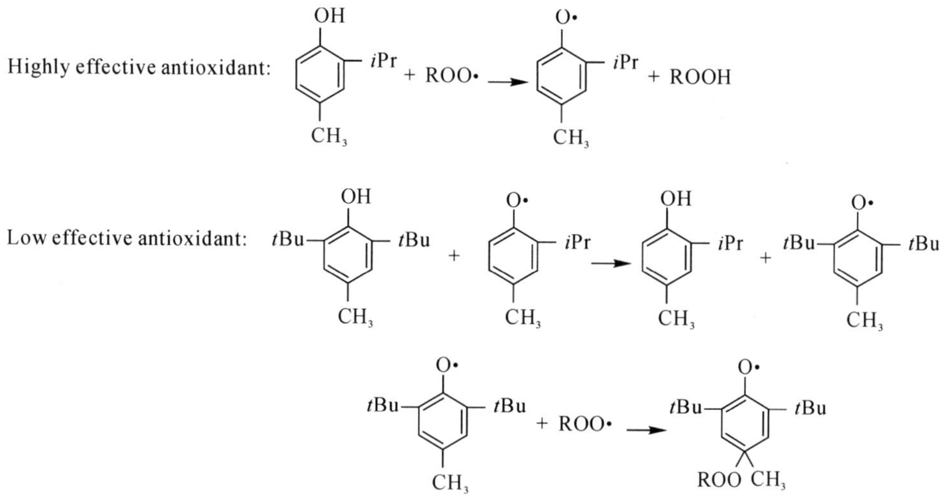

Fig. 2.21 Chemical process of the combination of phenol-type antioxidants

The combination of primary antioxidant and auxiliary antioxidant is heterogeneous synergetic effect, for example, the combination of phenol-type antioxidant and phosphite (see Fig. 2.22).

2. Additive effect

For the additive effect, there is no difference between separate use and combination use with other antioxidants.

3. Antagonistic effect (1+1<2)

The effect of combination of two or more antioxidants is worse than separately use is antagonistic effect. For example, to improve the thermal-oxidative stability as well as the photo-oxidative stability, frustrated phenol and organosulfur compounds are used together, which may generate synergetic effect. Further addition of frustrated amines can further improve the photo-oxidative stability. However, the acidity of oxidation products of organosulfur is very strong, which can convert amine to its ammonium salt, no further improvement of photo-oxidative properies can be achieved.

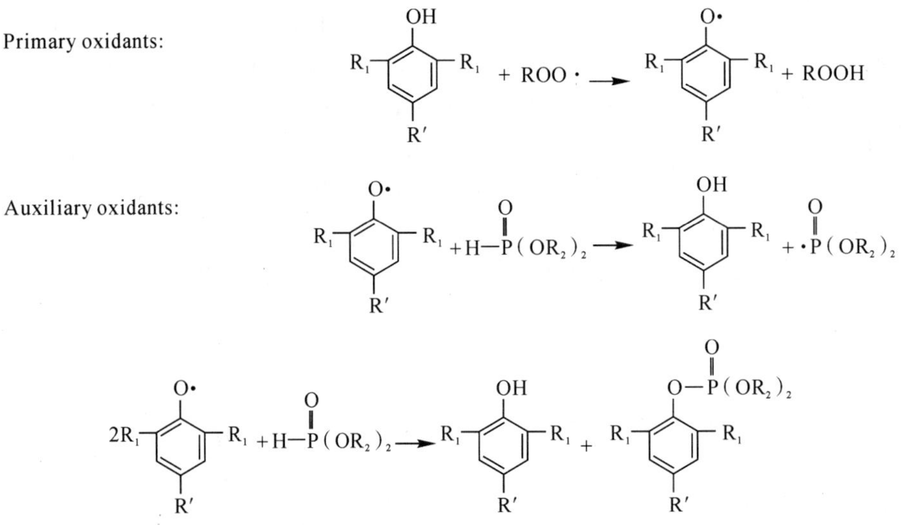

Fig. 2.22 Heterogeneous synergetic effect during the combination of phenol-type antioxidant and phosphite

2.4.4 Applications of thermal degradation

Although the word "degradation" usually has a pejorative connotation, a number of important industrial applications depend on this class of reactions to develop new products or processes.

1. Analytical Pyrolysis

Pyrolysis of a polymer means thermal degradation in the complete absence of any external reactant. Analytical pyrolysis, defined as pyrolysis conducted in combination with some physicochemical separation technique (pyrolysis-GC-MS, for instance), has found a wide range of applications in biopolymers and synthetic polymers.

Polymers pyrolyze by different combinations of three general mechanisms: random

Chapter 2 Thermal degradation and stabilization

scission (PE), depolymerization (POM, PMMA), or elimination of small molecules other than monomer (for instance, HCl in PVC). Each polymer under strictly controlled pyrolytic conditions provides a typical pattern of degradation products, which can be used for fingerprint identification, composition analysis, structure determination, or mechanistic studies.

2. Introduction of new chemical functionalities

Controlled thermal degradation of polyolefins in the presence of peroxides has been carried out in extruders or in continuous mixers, to reduce polymer MW and sample polydispersity. The resulting material has more processing advantages than the undegraded one. Based on the same principle, random chemical functionalization of polyolefins with polar groups has been achieved by extruding the polymer in the presence of peroxy ketals or peroxy esters.

3. Chemical modification of polymer structure

Carbon fiber is used in a variety of structural and electrical applications, and is probably the most well-known example of high-performance material produced by thermal degradation. Although carbon fiber can be produced in several ways, including alignment of molecules of pitch in its mesophase state, the most economical way consists of "hot-stretching" high MW polyacrylonitrile (PAN) fibers and while heating under controlled conditions. The filaments are first oxidized under tension at around 260 ℃ in air, to convert the PAN precursor to a thermally stable structure by an exothermic reaction, according to the following equation:

Stabilized PAN

The carbonization step is carried out in an inert atmosphere (high-purity N_2) above 1,000 ℃ to yield carbon fibers containing 93%–95% carbon and a tensile modulus (E) of 140–200 GPa. This grade is mostly used in sporting goods and in composites.

To achieve fibers in which the carbon crystals are further stretched and aligned, graphitization takes place around 2,000 – 3,000 ℃. These graphite fibers have a carbon content greater than 99% and a tensile modulus of 400 – 1,000 GPa. For comparison, the tensile modulus of carbon nanotubes is 1,000 GPa, and of diamond is 1,200 GPa.

4. Metal Injection Molding (MIM)

Polyacetals have a low ceiling temperature, and are readily depolymerized by unzipping at low temperature (0.4% – 0.8%/min^{-1} at 222 ℃ for POM). Owing to this low thermal stability, polyacetals can be used only if end-capped with stable groups (acetate or ether). This inherent thermal instability is exploited in an industrial method known as "metal injection molding", which allows fine metal powder mixed with a polymer binder to be processed by injection molding, in much the same way as thermoplastic materials. In a procedure based on POM, the binder is removed by thermal devolatilization according to following equation:

$$R-(CH_2-O-)_n H \xrightarrow{HNO_3/150\ ℃} R-CH_2OH + (n-1)H_2C=O$$

5. Recycling

Polymers can be recycled by reuse of existing material (primary recycling), by regranulating the waste by mechanical means so that it can be melted and formed again (secondary recycling), or by transforming the waste into new chemical compounds (tertiary recycling) through chemical reactions. The last of these methods presents several economic advantages in comparison to primary or secondary recycling, since revalorizing steps can be avoided. Recently, thermolysis has been viewed as a viable alternative to recovery for polymer recycling and numerous studies have been directed toward this objective.

The term "feedstock recycling" has been used to describe this new class of plastics recycling technology, which breaks down solid polymers into a spectrum of basic chemical compounds that can be reused as raw materials for the chemical industry. Vinyl polymers, when pyrolyzed at temperatures from 200 ℃ to 500 ℃ in the total absence of air, usually degrade to yield monomers (polymethyl methacrylate, poly α-methylstyrene), polystyrene, polyisobutylene) or a wide distribution of molecular fragments (polyethylene and polypropylene). In order to reduce the process temperature and to limit the range of products, particularly in the case of polyethylene, several catalyst cracking systems based on zeolites or clays have been developed.

Questions

1. Analysis the thermal degradation of PVA.
2. Analysis the thermal degradation of poly *tert*-butyl methacrylate.
3. How about the stability of graphene or graphene-like polymers?
4. How can BHT avoid the influence of Cu^{2+}?

5. Case study: compare the thermal degradation pathway of PMMA with different end groups (PMMA= CH, and PMMA—H).
6. Check literature and give at least two examples on the application of thermal stable polymer in aerospace.
7. Check literature and give at least two examples on the application of thermal stable polymer in daily life.

Further reading Ⅰ: thermal degradation

Industrial plastics are frequently exposed to elevated temperatures during melt processing and in many engineering applications. The glass transition temperature (T_g) and the melting point (T_m) of semicrystalline polymers are the most important thermal characteristics of polymers. These two parameters alone are generally not sufficient to predict the material's service temperature, if thermal stabilization is not properly controlled. Several polymer systems produce toxic low molecular weight compounds when subjected to heat, which can create hazardous conditions if they come into contact with a human body, either by ingestion or by breathing. Flammability of polymers used in construction is another concern when the polymers are exposed to high temperatures. All these factors combined make thermal stability appraisal a necessity in most application developments.

1. Thermal stability

The notion of thermal stability is a vague concept, since it depends primarily on the time scale of observation. A PMMA sample, for instance, can be stable at 300 ℃ for a few seconds, but can withstand only a temperature of 150 ℃ if the heating time spans several hours.

To be able to compare thermal stability between polymers of different structures, it is necessary to rely on some standardized systems, such as the temperature of half-decomposition ($T_{1/2}$). The temperature of half-decomposition is defined as the temperature at which the polymer loses half of its weight when heated in vacuo for 30 min. Experimentally, $T_{1/2}$ can be conveniently determined by thermal gravimetry (TG). From the TG curves obtained at different scan speeds, an Arrhenius plot at constant weight-loss ratio is derived. The pre-exponential factor and activation energy determined are then used to calculate $T_{1/2}$.

2. Polymer structure and thermal stability

Intrinsic chemical factors which influence heat resistance include primary bond strength, secondary or van der Waals interactions, hydrogen bonding, resonance stabilization, the mechanism of bond cleavage, structure regularity, intrachain rigidity, crosslinking, and branching. Owing to the presence of multiple secondary reactions, the effect of chemical structure on degradation kinetics is not easily rationalized from chemical first principles. The actual degradation kinetics is highly variable and depends not only on the polymer structures, but also on the reaction conditions: sample size, internal or external unstable structures, and additives. More specifically, it is experimentally observed that the degradation temperature and the product distribution can be controlled by changing the heating rate of the polymer, as a result of competitive pathways for degradation. Apart from the primary effects cited, differences in thermo-oxidative resistance can be discerned between different stereoisomers of the same compound (PMMA, PVC, and PP).

Not withstanding this complexity, some semi empirical rules could be identified by analogy with small organic "model" molecules, and by considering that thermal degradation is a homolytic bond scission process initiated by thermally activated molecular vibrations.

(1) The rate of thermal degradation is related to the number of pendent groups present on the polymer chain. Thus polyisobutylene (PIB) degrades faster than isotactic PP, which itself decomposes faster than HDPE.

(2) A rigid polymer backbone has less possibility of rearrangement and fragmentation and can withstand higher thermal energy.

(3) The heat resistance increases with the number of covalent bonds per repeat unit: a crosslinked or ladder polymer can be broken only after scission of two or more covalent bonds.

(4) The bond of lowest energy is the first bond to be cleaved. Conversely, polymers with multiple bonds and aromatic structures are less prone to thermal degradation.

Poly *para*-phenylene(PPP), with its rodlike structure composed of highly delocalized π-electron orbitals, satisfies most of the requirements for high thermal stability. Possessing $T_{1/2} > 400$ °C and an estimated melting point of 1,400 °C, PPP constitutes a reference for heat-resistant polymer. PPP can be synthesized by a number of routes, with the best heat-resistant material obtained by polymerization of 1,3-cyclohexadiene, followed by dehydrogenation of the polymer formed according to the following equation:

$$n \bigcirc \longrightarrow \left[\bigcirc\right]_n \xrightarrow{-H} \left[\bigcirc\right]_n$$

Moldable polymers can be obtained by increasing chain flexibility with substitution of lateral groups, at the expense of decreasing thermal stability. Polyimides and polyetherimides, with a semi rigid structure and electronically stabilized aromatic rings, form an important class of melt-processible polymers with outstanding heat resistance.

Apart from the decomposition temperature, the propensity for char formation is another polymer thermal characteristic of practical importance. The char-forming tendency generally shows a negative correlation with flammability: a sample with the greatest char residue is also the least flammable polymer and the presence of flame retardant additives also increases char formation. In the case of fire, the amorphous crust of char material that forms on the fluid surface reduces the heat flux to the burning fluid, which in turn reduces the combustion rate.

Further reading II : auto-accelerated oxidation of plastics

The thermo-oxidative degradation of polyolefins and many other plastics is usually an auto-accelerated process, that is, the rate of oxidation is low or negligible in the beginning but steadily increases and often reaches a constant rate of oxidation.

Like ordinary chain reactions, the overall oxidation process can be divided into three stages: called initiation, propagation, and termination of degradation. The first stage, the so-called initiation, is the formation of free radicals via C—C and/or C—H bond cleavage which may be induced by heat, UV light, mechanical stress or by trace amounts of initiators such as peroxides and/or hydroperoxides.

$$R-H \rightarrow R\cdot + H\cdot$$
$$R-R \rightarrow R\cdot + R\cdot$$

Alkyl radicals can undergo a number of "chain" reactions. Propagation of decomposition usually starts when (atmospheric) oxygen reacts with the newly formed free chain radicals $R\cdot$. This reaction produces highly reactive peroxy radicals $ROO\cdot$ which then react with labile hydrogen of polymer chains to form more free radicals $R\cdot$ and hydro peroxides ROOH.

$$R\cdot + \cdot O-O\cdot \rightarrow ROO\cdot$$
$$ROO\cdot + RH \rightarrow R\cdot + ROOH$$

The second reaction involves breaking C—H bonds which requires considerable energy unless the polymer contains reactive hydrogen. Thus, this reaction is often the rate determining step.

The ROOH can decompose via homolytic cleavage to alkoxy and hydroxy radicals which, in turn, also abstract labile hydrogen from polymers which leads to the formation of even more free chain radicals.

$$ROOH \rightarrow RO\cdot + \cdot OH$$
$$RO\cdot + RH \rightarrow R\cdot + ROH$$
$$\cdot OH + RH \rightarrow R\cdot + H_2O$$

However, the cleavage of an O—O bond has a rather high activation energy. For example, the dissociation energy of hydroperoxide is in the range of 200 kJ/mol. For this reason, a bimolecular decomposition mechanism with lower activation energy is more likely to occur.

$$ROOH + RH \rightarrow RO\cdot + R\cdot + H_2O$$

The autocatalytic decomposition reactions are usually very fast when compared to (heat induced) initiation reactions (C—C bond cleavage). Since the concentration of hydroperoxide is very low at the beginning of an autooxidation process, the rate of decomposition is very low but steadily increases until the hydroperoxide concentration reaches a steady state which means the rate of peroxide decomposition is equal to that of formation (see Fig. 2.23).

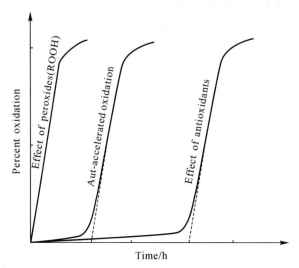

Fig. 2.23 The rate of different kinds of oxidation

Further reading Ⅲ : effects of peroxides and antioxidants on auto-oxidation of plastics

It has been postulated that the bimolecular decomposition of hydroperoxides is the rate determining step for thermo-oxidative decomposition. However, this is not always true since even trace amounts of impurities such as transition metals can have a large impact on the oxidative stability of plastics. Particularly, Co, Cu, Fe, Mn can greatly accelerate oxidation because they are potent catalysts for hydroperoxide decomposition and thus, greatly reduce the activation energy.

$$ROOH + M^+ \rightarrow RO\cdot + HO^- + M^{2+}$$
$$ROOH + M^{2+} \rightarrow ROO\cdot + H^+ + M^+$$

Unsaturated polymers such as polyisoprene and polybutadiene are most susceptible to metal-catalyzed oxidation because double bonds markedly decrease the activation energy of oxidation reactions. Branching and crystallinity also affect the thermo-oxidative stability of plastics. In general, polymers with high crystallinity and no branching are more stable than amorphous polymers of low density and low crystallinity, such as low density polyethylene. The general order of stability is listed below:

$$HDPE > LLDPE > LDPE > i\text{-}PP > PS > PBD > PI$$

To prevent or at least to slow down oxidative chain reactions, antioxidants are usually added. Typically a combination of sterically hindered phenols and organic phosphites are added. The phosphites decompose the hydroperxides into harmless inert products and the hindered phenols scavenge the free radicals.

Chapter 2 Thermal degradation and stabilization

The oxidation reactions stop when either two radicals recombine (dimerization or cross-linking) or when a chain radical abstracts a hydrogen from another polymer chain (disproportionation/hydrogen abstraction). These reactions always occur but can be accelerated by addition of stabilizers. Recombination of two chain radicals results in an increase of the molecular weight and crosslinking density.

$$R\cdot + R\cdot \rightarrow R\text{-}R$$
$$2ROO\cdot \rightarrow ROOR + O_2$$
$$RO\cdot + RO\cdot \rightarrow ROOR$$
$$R\cdot + RO\cdot \rightarrow ROR$$
$$HO\cdot + ROO\cdot \rightarrow ROH + O_2$$

These reactions cause embrittlement and cracking of the plastics. Termination by chain scission, on the other hand, decreases the molecular weight and thus, softens the plastic.

$$R_n\cdot + R_m\cdot \rightarrow R_{n-2}\text{—}CH=CH_2 + R_m$$
$$2RCOO\cdot \rightarrow RC=O + ROH + O_2$$

Both chain scission and chain hardening has a negative impact on the mechanical properties. Which of these termination reactions dominates depends on the type of plastic and on the reaction conditions. For example, polyolefins with short alkyl side groups like polypropylene and polybutylene, and unsaturated polymers like natural rubber (polyisoprene) undergo predominantly chain scission, whereas polyethylene, and rubbers with somewhat less active double bonds like polybutadiene and polychloropene become brittle due to crosslinking.

Further reading IV: thermal-oxidative degradation of rubbers

Most elastomers will undergo significant changes over time when exposed to heat, light, or oxygen (ozone). These changes can have a dramatic effect on the service life and properties of the elastomers and can only be prevented or slowed down by the addition of UV stabilizers, antiozonates, and antioxidants.

Depending on the microstructure of the diene elastomer, oxidative degradation will either cause hardening or softening. For example, polybutadiene usually undergoes oxidative hardening whereas polyisoprene softens when exposed to heat and oxygen. Hardening is much more common because free radicals produced when rubber is exposed to heat, oxygen and/or light rapidly combine and in this process form new crosslinks. This drastically reduces the flexibility of the rubber. Polymers with pendent bulky side groups, on the other hand, will undergo strain softening because radical recombination reactions are less likely to occur due to steric hindrance. Instead, these polymers degrade by chain scission caused by disproportionation and hydrogen abstraction.

The aging of a rubber due to oxidation and heat is greatly accelerated by stress, and exposure to other reactive gases like ozone. Besides embrittlement (chain hardening) or softening (chain scission) other visible changes such as cracking, charring, and color fading is observed.

Although the general mechanism of autooxidation is well understood, the actual chain scission and crosslinking steps are often unknown. They depend on the composition of the rubber including concentration of accelerators, activators, and fillers as well as on the temperature and composition of the atmosphere. Two possible mechanisms of thermal oxidation with subsequent chain scission or crosslinking are shown below. The process is very complex and involves several intermediates and side reactions.

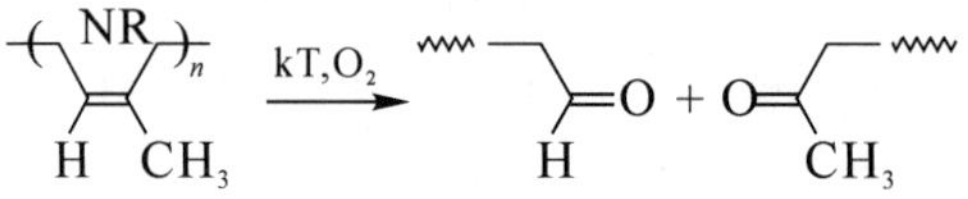

A. Oxidative softening-chain scission

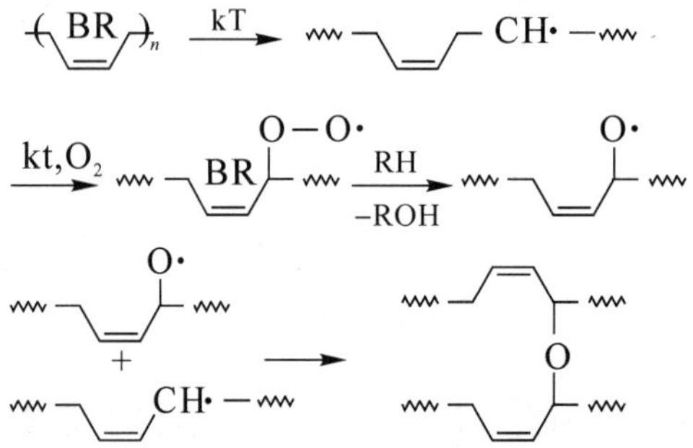

B. Oxidative hardening-crosslinking

Further reading V : general mechanism of thermal degradation

In general, the type of degradation (chain hardening or softening) depends on the chemical composition of the polymer. For example, crosslinking dominates in polybutadiene and its copolymers such as polybutadiene (BR), styrene-butadiene-styrene (SBS), acrylonitrile butadiene(NBR) and in many diene rubbers with less active double bonds due to electron-withdrawing groups such as halogen [e. g. polychloroprene rubber(CR)] whereas elastomers with bulky and/or electron donating side groups ($-CH_3$) attached to the carbon atom adjacent to the double bonds are vulnerable to chain scission. This includes natural rubber (NR), polyisoprene rubber(IR), isobutylene isoprene rubber (IIR), and any other unsaturated polymer with electron donating groups. Some other polymers such as styrene butadiene(SBR), ethylene-propylene monomer (EPM), and EPDM undergo both chain

Chapter 2 Thermal degradation and stabilization

scission and crosslinking. However, often crosslinking reactions dominate so that these rubbers harden over time.

The resistance to oxidative degradation depends on many factors, including chemical composition, molecular weight, crosslink density, and type of cross-links. Diene elastomers that have electron-donating groups attached to the diene are usually the least stable rubbers (e.g. NR, IR). They have poor heat, ozone and UV resistance, whereas elastomers with a low number of double bonds [hydrogenated nitrile rubber(HNBR), IIR, EPDM] have good or even excellent heat resistance.

Table 2.5 shows the degradazion of different ruber.

Table 2.5 Degradation of different rubber

Rubber	Chemical structure	Type of degradation
Natural rubber(NR)	$\{CH_2-CH=C(CH_3)-CH_2\}_n$	Chain scission (softens)
Polyisoprene rubber(IR)	$\{CH_2-CH=C(CH_3)-CH_2\}_n$	Chain scission (softens)
Polychloroprene rubber(CR)	$\{CH_2-C(Cl)=CH-CH_2\}_n$	Cross-linking & chain scission (hardens)
Polybutadiene rubber (BR)	$\{CH_2-CH=CH-CH_2\}_n$	Cross-linking (hardens)
Styrene butadiene (SBR)	$\{CH_2-CH=CH-CH_2\}_n\{CH_2-CH(C_6H_5)\}_m$	Cross-linking & chain scission (hardens)
Acrylonitrile butadiene rubber(NBR)	$\{CH_2-CH=CH-CH_2\}_n\{CH_2-CH(C\equiv N)\}_m$	Cross-linking (hardens)
Isobutylene Isoprene rubber(IIR)	$\{CH_2-CH=CH-CH_2\}_n\{C(CH_3)_2-CH_2\}_m$	Chain scission (softens)

The stability of an elastomer is also affected by other ingredients in the rubber compounding formulation. For example, under certain conditions (residual) cross-linkers and accelerators confined in the elastomer can decrease the thermal stability because they easily undergo thermal decomposition at elevated temperature producing radicals that are capable of accelerating thermo-oxidative degradation of the network. Soluble fatty acid salts of metal ions such as Cu, Mn, Ni, Co, and Fe act as catalysts for oxidation, and thus greatly accelerate the thermo-oxidative decomposition of rubber.

The composition of the atmosphere plays an equally important role. For example, ozone, even when present in very small concentration, will cause extensive cracking perpendicular to the stress in the rubber.

Further reading Ⅵ: antioxidants

Polymers will change over time when exposed to radiation, excessive heat and/or corrosive environments. These changes are the result of oxidative degradation caused by free radicals which form through hydrogen abstraction or homolytic scission of carbon-carbon bonds when polymers are exposed to heat, oxygen, ozone, or light. These changes can have a dramatic effect on the service life and properties of the polymer.

To prevent or slow down degradation, antioxidants and UV stabilizers are often added. The two main classes of antioxidants are free-radical scavengers and peroxide scavengers. The free-radical scavengers are sometimes called *primary antioxidants* or radical chain terminators whereas peroxide scavengers are often called *secondary antioxidants* or hydroperoxide decomposers.

1. Primary antioxidants (free-radical scavengers)

As the name suggests, free-radical scavengers react with chain-propagating radicals such as peroxy, alkoxy, and hydroxy radicals in a chain terminating reaction. To be more specific, these antioxidants donate hydrogen to the alkoxy and hydroxy radicals which converts them into inert alcohols and water respectively.

Typical commercial primary antioxidants are hindered phenols and secondary aromatic amines. These compounds come in a wide range of molecular weights, structures, and functionalities.

The most widely used primary antioxidants are sterically hindered phenols. They are very effective radical scanvengers during both processing and long-term thermal aging, and are generally non-discoloring. Many also have received FDA approval. The mechanism of scavenging oxy radicals is shown below:

$$\text{Ar-OH} \xrightarrow[\text{ROOH}]{\text{ROO·}} \text{Ar-O·} \xrightarrow{\text{ROO·}} \text{Ar-O·(O-O-R)}$$

The most effective primary antioxidants are secondary aromatic amines. However, they cause noticeable discoloation and can only be used if discolation is not a problem, like carbon filled rubber products. They also function as antiozonants and metal ion deactivators.

2. Secondary antioxidants (peroxide scavengers)

As the name suggests, peroxide scavengers (secondary antioxidants) decompose hydroperoxides (ROOH) into nonreactive products before they decompose into alkoxy and hydroxy radicals. They are often used in combination with free radical scavengers (primary antioxidants) to achieve a synergistic inhibition effect.

The most common secondary antioxidants are trivalent phosphorus compounds (phosphites). They reduce hydroperoxides to the corresponding alcohols and are themselves

Chapter 2 Thermal degradation and stabilization

transformed into phospates. The general mechanism of peroxide decomposition is shown below:

$$RO-\underset{\underset{RO}{|}}{\overset{\overset{RO}{|}}{P}} + ROOH \longrightarrow ROH + RO-\underset{\underset{RO}{|}}{\overset{\overset{RO}{|}}{P}}=O$$

Another class of secondary antioxidants are thioethers or organic sulfides. They decompose two molecules of hydroperoxide into the corresponding alcohols and are transformed to sulfoxides and sulfones.

$$ROOH + R-S-R \longrightarrow ROH + R-\overset{\overset{O}{\|}}{S}-R$$

$$ROOH + R-\overset{\overset{O}{\|}}{S}-R \longrightarrow ROH + R-\underset{\underset{O}{\|}}{\overset{\overset{O}{\|}}{S}}-R$$

Organic sulfides are very effective hydrogen peroxide decomposers during long-term thermal aging and are often used in combination with other antioxidants that provide good protection during processing, like hindered phenols.

In order to choose the most effective stabilizer package, it is important to know what temperature range the polymer will be exposed to. A good stabilizer package should protect the plastic during both processing, where high temperatures are encountered to melt and form the resin, and during lifetime when exposed to its upper service temperature.

Further reading Ⅶ: sterically hindered phenolic antioxidants

Some of the most widely used antioxidants are sterically hindered phenols. These compounds act as a primary antioxidants by converting peroxyl radicals to hydroperoxides. Thus, they inhibit autooxidation of organic polymers by reducing the amount of peroxyl radicals. The mechanism of scavenging peroxy radicals is shown below for butylated hydroxytoluene (2,6-*di-tert*-butyl-4-methylphenol, BHT):

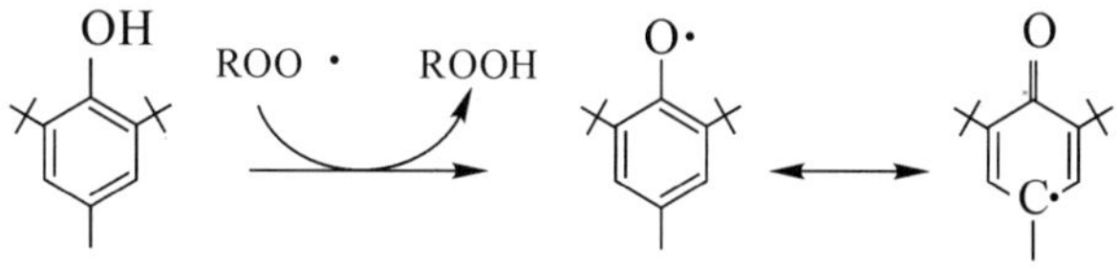

The peroxyl radicals abstract the reactive hydrogen of the BHT. The resulting oxytoluene radicals are stabilized by electron delocalization and by the bulky substituents (steric hindrance) which greatly reduces their reactivity and thus prevents them from reacting with polymer chains (hydrogen abstraction). Instead, the oxytoluene radicals undergo a number of other reactions. For example, they can undergo a bimolecular reaction that produces another BHT molecule and a *p*-quinomethane or the oxytoluene radical can undergo an irreversible reaction with a second peroxyl radical which thermally decomposes to *para*-quinone.

Thus, each BHT molecule can react with two peroxyl radicals to yield products that are very stable to further reactions. Recombination of two oxytoluene radicals is also possible. Thus, all oxytoluene radicals are ultimately transformed to quinonoid structures.

Butylated hydroxytoluene (BHT) is only one among many sterically hindered phenols. Other represenatives include butylated hydroxyanisole (BHA), *tert*-butylhydroquinone (TBHQ) and gallates among many others which all undergo similar reactions.

TBHQ   2-BHA   3-BHA   Gallate

Chapter 3 Photo degradation and stabilization

Photo degradation is induced by light, which is very common in the application of polymeric products, especially sunlight and high energy irradiation. In case of aerospace application, photo degradation under high energy irradiation is also necessary to be studied.

3.1 Photo degradation

3.1.1 Basics of photo process

After absorption of irradiation such as UV light by polymers, they will go to corresponding excited state. The excited molecules may undergo different process: photophysics process and photochemistry process.

As shown in Fig. 3.1, in the photophysics process, the absorbed photo energy will be converted to either irradiation with longer wavelength or heat, and the excited molecule will go back to its ground state; in the process of photochemistry, some chemical reactions will happen at the excited state to yield subsequent photo degradation of polymer. In presence of O_2, the corresponding photo degradation is photo-oxidative degradation.

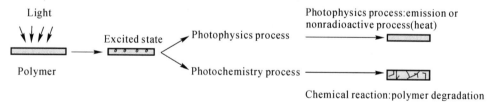

Fig.3.1 Photophysics and photochemistry process of polymer

1. Photophysics process

As demonstrated in Jablonski diagram (see Fig. 3.2), S_0 is the ground state before molecule is excited; singlet state S_1, S_2, S_3, $\cdots$, S_n is the excited states after molecules are excited. It should be mentioned that photochemical reactions is from the lowest excited singlet state S_1 due to the short lifetime of higher excited states. After being excited to its excited state, the excited molecule will undergo irradiation or non-irradiation process to lose its energy. For the non-irradiation process, the excited state will go back to ground states through vibration to generate heat; for the irradiation process, fluorescence (S_1 to S_0) with lifetime of $10^{-6} \sim 10^{-9}$ s and phosphorescence (T_1 to S_0) with lifetime of ca. 10^{-3} s will be generated. In general, the photophysical process is as follows:

(1) Photon #1 hits molecule.

(2) Photon #1 absorbed, electron at ground state gains energy and jumps to higher level.

(3) Electron loses some of that energy.

(4) Electron jumps back down to ground state and emits photon #2.

(5) Photon #2 has less energy, corresponding to a different color (wavelength of light).

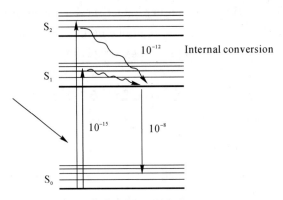

Fig.3.2 Jablonski diagram

Therefore, the photophysical process is wavelength dependent.

2. Photochemistry process

Because the number of excited states that can involve in photochemical reaction is limited, the photochemistry process is independent on wavelength of the absorbed light. No matter which wavelength of light is absorbed, they will go to the same excited states (S_1 and T_1), the excess energy will be released through vibration and other ways.

Normally, after the absorption of photo, the following reactions may happen (see Fig. 3. 3): the first process is not a chemical process, it can generate light, belonging to photophysical process; the others are photochemical processes, in which an active intermediate M* is involved. M* will transfer its energy to molecule A at ground state, while M* is converted to M, A is excited to A*. Some chemical reactions may happen for A* to generate new compound B. In the process, M* is a photosensitizer.

$$M^* \longrightarrow M + h\nu$$
$$M^* \longrightarrow E$$
$$M^* \longrightarrow M点(或M^+, M^-) \longrightarrow E$$
$$M^* + A \longrightarrow M^+A^- \longrightarrow E$$
$$M^* + A \longrightarrow M + A^*$$
$$A^* \longrightarrow B$$

Fig.3.3 Possible photoreactions

Chapter 3 Photo degradation and stabilization

For example, ketone and related molecules will be photodecomposed to radicals upon absorption of light:

$$R-\underset{O}{\underset{\|}{C}}-R_1 \xrightarrow{h\nu} R-\underset{O}{\underset{\|}{C}}\cdot + R_1\cdot$$

For example, aldehyde and related molecules will be photo decomposed to new molecules:

$$R-\underset{O}{\underset{\|}{C}}-H \xrightarrow{h\nu} R-H + CO$$

Usually, not all of the absorbed photo will be used in the photochemical process. Einstein introduced the concept of quantum yield (φ):

$\varphi = $ the number of the emitted quanta/the number of the absorbed quanta

The number of quantum yield is very broad, ranging from very small to millions. We can use φ to determine the type of chemical reaction. If $\varphi < 1$, the photochemical reaction is a direct reaction; if $\varphi > 1$, the reaction is a chain reaction.

3. Absorption of UV radiation by polymers

Photo-oxidation is the most common mode of weathering for industrial polymers, and differs from thermal degradation principally in the mode of chemical activation. In place of vibrational energy, the energy of the photon provides the driving mechanism for free-radical generation. Photochemical processes are based on two fundamental principles: the Grotthus-Draper law, which states that only radiant energy that is effectively absorbed can activate molecules, and the Stark-Einstein law, which asserts that one absorbed photon can induce photochemical reaction of only a single molecule. The first law implies the presence of an appropriate chromophore in the polymer, whereas the second law indicates that the formation of photoproducts is linearly dependent on the light intensity, and that a photoreaction can occur only if the energy of the photon is sufficient to overcome the corresponding activation energy.

For most organic molecules, absorption in the near-UV (190 – 400 nm) involves electronic transition from the highest occupied molecular orbital (HOMO) to the lowest unoccupied molecular orbital (LUMO). Commercial polymers can be broadly divided into two categories, depending on whether or not they contain delocalized p-electrons, sometimes combined with heteroatoms with nonbonding valence shell electron pairs (O, N, S) in their chemical structures (see Table 3.1). Polymers which contain carbonyls (C=O), conjugated polyenes (C=C—C=C), ketenes (C=C—C=O), and sulfones (—SO$_2$—) belong to the first group of chromophores and are capable of absorbing light in the near UV (190 nm < λ_{max} < 400 nm). Those from the second group possess only single covalent bonds, such as C—C, C—H, C—O, C—Cl, and C—F, and can in principle absorb light only in the far UV (λ_{max} < 190 nm).

Owing to vibrational and electronic couplings, UV absorption bands of polyatomic molecules in the condensed state are generally quite broad, with width at half-maximum commonly exceeding 60 – 80 nm. Even with this broadness, polymers such as polyolefins, polyacetals, polyvinyl chloride, or polyacrylonitrile should neither absorb nor degrade when exposed to light with wavelength above 230 nm. In practice, however, the UV absorption spectra of commercial samples of all the cited polymers show a broad absorbance band which extends well above the expected limits. Generally, these extraneous absorbances are weak and can be determined only after proper signal processing. This situation is illustrated in Fig. 3.4 for commercial additive-free films of HDPE and i-PP.

Table 3.1 Principal chromophore groups in synthetic polymers[a]

Polymer	Chromophore	λ_{max}/nm	ε_{max}[L/(mol/cm)]
Polyesters	>C=O	188(279)	900(15)
Polyaromatics	—φ—	200(256)	4,400(226)
Polyaryl ketones	—φ—CO—φ—	250(350)	18,000(120)
Polydienes	>C=C<	185(230)	8,000(2)
Conjugated polyenes	>C=C—C=C<	217	20,900
	—(C=C)$_3$—	263	~5×10^4
	—(C=C)$_{10}$—	432	~2×10^5
Ketenes	>C=C—C=O	220(350)	2×10^4(30)
Sulfones	—SO$_2$—		

[a] Values in brackets refer to the secondary absorption band.

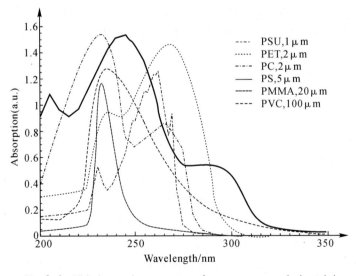

Fig.3.4 UV absorption spectra of some common industrial polymer films, at the thickness indicated

Far-UV spectroscopy indicates that the "absorption edge" of long-chain alkanes starts at about 155 nm (see Fig. 3.5). Absorption at much longer wavelengths observed in commercial polyolefins could be explained only by the presence of chromophore impurities and chemical defects formed during the synthesis, storage, and processing of the polymer. The absorption peak at 180 nm recorded in HDPE, for instance, has been attributed to vinylidene end groups, as revealed by independent ultraviolet-visible spectroscopy(UV-Vis) measurements.

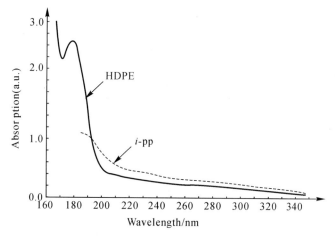

Fig.3.5 UV absorption spectra of 50 mm-thick films of additive-free commercial HDPE (—) and i-PP (·······)

Impurities in polymers can be classified as internal or external.

(1) Internal impurities. These are part of the molecular structures, situated either along the chain or at chain end(s), and may consist of:

1) Anomalous structural units, which result from the kinetics of polymerization.

2) In-chain peroxides, formed during polymer synthesis. Free-radical polymerization is generally accomplished without strict exclusion of air and small amounts of oxygen dissolved in the monomer will be scavenged by the macroradicals and included in the polymer chain as peroxide linkages. Photolysis of these peroxide groups gives alkoxy radicals, leading ultimately to hydroperoxides. Alternatively, if polymerization has been carried at elevated temperatures, the peroxides incorporated will undergo thermal fragmentation and disproportionation to form phenyl alkyl ketone end-group chromophores.

3) Carbonyl containing groups formed during material transformation. Processing, with elevated temperature and high shearing stresses, provides ample opportunities for thermal oxidation. Even at ambient temperature, some polymers such as isotactic polypropylene or cis-polybutadiene, can be readily oxidized if unstabilized.

(2) External impurities. Such impurities are contained in the sample but not

incorporated in the polymer structures. During synthesis, processing and storage, the polymer can be contaminated or blended with a variety of external chemical species which may contain chromophore and photoreactive groups. Some typical external impurities found in commercial plastics are:

1) Residual catalysts and initiators.

2) Traces of solvents (aromatic solvents, in minute amounts, can act as photosensitizers).

3) Pigments, dyes, and additives, traces of metal, metal oxides, or metal salts from the reactors, processing equipment, and containers.

3.1.2 Photo degradation and photo crosslink

Typically, photo degradation is a radical process. Depending on formation of radicals and their further reactions, there are two types of photo degradations: randomly degradation and depolymerization.

1. Randomly degradation

In this kind of photo degradation, randomly formation of radicals at any position of polymer chain is the source of degradation. Small species that larger than monomer unit will form during this process. The formed radical can be involved in complicate reactions, which also include crosslink reaction. Therefore, the degradation competes with the crosslink process. If the degradation is dominant over crosslink, it is randomly degradation.

$$\sim\sim M-M-M-M-M-M \sim\sim \xrightarrow{h\nu} M-M-M\cdot + \cdot M-M-M-M-M \sim\sim$$

If photo irradiation can break all backbone, and all the bond energy of the backbone are the same, we can get that:

$$\frac{dn_p}{dt} = \varphi \cdot I_a$$

Where n_p is number of molecules at degree of degradation $= p$, t is irradiation time, I_a is intensity of light, φ is quantum yield, M is weight of polymer, M_r is molecular weight of monomer.

The degree of photo degradation can be calculated:

$$p = \frac{M_r}{m \cdot N} \varphi \cdot I_a \cdot t$$

We can conclude that the degree of photo degradation is in proportion to I_a, t and φ. If the number averaged degree of polymerization is $\bar{u}_0$, then, at the degree of photo degradation of p, the corresponding number averaged degree of polymerization $\bar{u}_p$ should have the following relation:

$$\frac{1}{\bar{u}_p} - \frac{1}{\bar{u}_0} = p = \frac{M_r}{m \cdot N} \varphi \cdot I_a \cdot t$$

Thus, the quantum yield can be calculated from the slope of the following plots (see Fig. 3.6).

Chapter 3 Photo degradation and stabilization

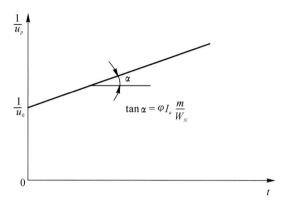

Fig.3.6 Relationship between molecular weight and time

2. Depolymerization

Depolymerization can be regarded as the reverse process of polymerization, it is a chain process. Once a radical is generated at the end or any position of the polymer chain, depolymerization will happen by lose of monomer one by one. This process is very similar to the depolymerization (unzipping reaction) in thermal degradation. The following elementary reactions will be involved:

(1) Initiation. The initiation can start from the end of the chain, the reaction rate can be calculated:

$$-\frac{dn_p}{dt}=k_1 n_p$$

where n_p is the number of molecules with photo degradation degree of p. k_1 is the function of I_a and φ, I_a is the intensity of light absorbed by the polymeric products, φ is the quantum yield.

The initiation can also start from any position of the polymer chain, then, the reaction rate can be calculated:

$$-\frac{dn_p}{dt}=k_1 n_p \cdot \bar{u}$$

Where $\bar{u}$ is the average degree of polymerization (DP).

(2) Depolymerization. The depolymerization rate is much faster than the initiation reaction, which can start at the end of polymer chain or any position of the polymer chain. Compare with the total amount of the depolymerized chains, their concentration is very small.

(3) Termination. Termination reaction is a first order reaction. Termination may happen between radicals and solvent, stabilizer, also including transfer induced deactivation.

3. Quantum yield of photo degradation

The quantum yield of the chain cleavage can be calculated according to the following equation:

$$\varphi = \frac{m \cdot N}{M_t I_B \cdot t} p$$

We can introduce the limit viscosity formula ($[\eta] = \lim_{\tau=0}(\eta \text{ sp}/c) = \lim_{\tau=0}(\ln \eta r/c)$ into the above equation to give:

$$\rho_\varphi = \frac{\rho N}{M_r} \frac{\{[\eta]_0/[\eta]_p\}^{1/\alpha} - 1}{I_n \cdot t}$$

Where ρ is the concentration of the polymer solution (g/100 mL), M_{n_0} is molecular weight of number of molecules of n_0 before photo degradation.

3.2 Photo-oxidative degradation

The mechanism of photo-oxidative degradation is very similar to thermal-oxidation degradation, including initiation, propagation and termination. The only difference is the initiation process of photo-oxidation degradation is light, which is more common than thermal-oxidative degradation. Meanwhile, the propagation process is also affected by light.

3.2.1 Elementary reaction of photo-oxidative degradation

There are some elementary reactions for photo-oxidative degradation, such as initiation, propagation and termination.

1. Initiation

The polymers will form excited state by irradiated by light or photo to form radical. This radical can further react with oxygen to form ROO·:

$$RH \xrightarrow{h\nu} RH* \longrightarrow R\cdot + H\cdot$$
$$R\cdot + O_2 \longrightarrow ROO\cdot$$

2. Propagation

Compare with the propagation reaction of thermal-oxidative degradation, the bond of RO—OH and R—OOH can be cleaved by light to generate more radicals. For example, the dissociation energy of RO—OH and R—OOH is 175.8 kJ/mol and 293 kJ/mol, respectively, which can be cleaved by UV light with wavelength of 300 nm; however, the energy of 300 nm UV light cannot break the bond of ROO—H, whose dissociation energy is ca. 376.8 kJ/mol.

ROO· + RH ⟶ ROOH + R·
ROOH ⟶ R· + ·OOH

R· + O_2 ⟶ ROOH·
ROO· + RH ⟶ $\boxed{ROOH}$ + R·
$\boxed{ROOH}$ ⟶ RO· + HO·

ROOH ⟶ RO· + ·OH
RO· + RH ⟶ ROH + R·
HO· + RH ⟶ R· + H_2O
$\boxed{2ROOH}$ ⟶ RO· + ROO· + H_2O
RO· + RH ⟶ R· + ROH
HO· + RH ⟶ R· + H_2O

Chapter 3 Photo degradation and stabilization

3. Termination

At higher concentration of oxygen, the termination reaction follows the bimolecular termination of $ROO\cdot$; while at lower concentration of oxygen, the termination reaction is the termination of $ROO\cdot$ and $R\cdot$, while the termination between $ROO\cdot$ is also present.

$$High[O_2]: ROO\cdot + ROO\cdot \longrightarrow ROOR + O_2$$
$$Low[O_2]: ROO\cdot + R\cdot \longrightarrow ROOR$$
$$R\cdot + R\cdot \longrightarrow R-R$$

If two $ROO\cdot$ is closed to each other, they can react with each other to form stable cyclized peroxide or epoxy.

$$\sim\sim CH_2-\underset{\underset{O\cdot}{\underset{|}{O}}}{\overset{R}{\underset{|}{C}}}-CH_2-\underset{\underset{O\cdot}{\underset{|}{O}}}{\overset{R}{\underset{|}{C}}}-CH_2\sim\sim \longrightarrow \sim\sim CH_2-\underset{\underset{O-}{\underset{|}{O}}}{\overset{R}{\underset{|}{C}}}-CH_2-\underset{\underset{-O}{\underset{|}{}}}{\overset{R}{\underset{|}{C}}}-CH_2\sim\sim + O_2$$

$$\sim\sim CH_2-\underset{\underset{\underset{O\cdot}{\underset{|}{O}}}{\underset{|}{O}}}{\overset{R}{\underset{|}{C}}}-\underset{\underset{O\cdot}{\underset{|}{}}}{CH}-CH\sim\sim \longrightarrow \sim\sim CH_2-\underset{\underset{O}{\diagup\diagdown}}{\overset{R}{\underset{|}{C}}}-CH-\overset{R}{\underset{|}{C}}H\sim\sim + O_2$$

Meanwhile, these polymeric radicals can react with each other, as well as to react with polymeric $ROO\cdot$ to crosslink. The cleavage and crosslink reactions can happen simultaneously, while the cleavage will decrease the molecular weight, while crosslink will increase the molecular weight to make the polymer brittle.

3.2.2 Source of photo-oxidative degradation

Generally, polymers with double bonds are easy to be excited after the absorption of UV light. For polymers with only single bond, it should be very stable because it will not absorb any UV light. The impurities in polymers, such as trace amounts of catalyst, additives, metal ions as well as trace amounts of peroxide and carbonyl groups may also induce photo degradation or photo-oxidative degradation.

1. Metal ions

Metal ions are sensitizer for the photo-oxidative degradations of polyolefin. For example, the trace amounts of Ti left by Zieglar-Natta catalyst is the main source for the photo-oxidative degradation of polyolefin. During the polymerization and processing of polymer, it is possible to introduce trace amounts of Fe, which is also the source of photo-oxidative degradation.

2. ROOH

ROOH can be generated from thermal-oxidative degradation and the very beginning of photo-oxidative degradation. ROOH can absorb light with wavelength of ca. 210 nm, even to 300 nm, which can break the bond of O—O:

$$ROOH \xrightarrow{h\nu} RO\cdot + \cdot OH$$

The quantum yield of this reaction is close to 1. The formed macromolecular radical ($RO\cdot$) can abstract protons from polymer to form $R\cdot$. Such radical can react with oxygen to form $ROO\cdot$. Meanwhile, the $RO\cdot$ radical can also undergo β-cleavage to form ketone and alkyl radical ($R\cdot$).

Regeneration of $R\cdot$:

$$\sim\sim CH_2-\underset{\underset{O\cdot}{|}}{\overset{\overset{CH_3}{|}}{C}}-CH_2\sim\sim + RH \longrightarrow \sim\sim CH_2-\underset{\underset{OH}{|}}{\overset{\overset{CH_3}{|}}{C}}-CH_2\sim\sim + R\cdot$$

$$R\cdot \xrightarrow{O_2} ROO\cdot$$

Formation of ketone:

$$\sim\sim CH_2-\underset{\underset{O\cdot}{|}}{\overset{\overset{CH_3}{|}}{C}}-CH_2\sim\sim \longrightarrow \begin{cases} \sim\sim CH_2-\overset{\overset{O}{\|}}{C}-CH_2\sim\sim + CH_3\cdot \\ \sim\sim CH_2-\overset{\overset{O}{\|}}{C}-CH_2\sim\sim + \cdot CH_2\sim\sim \end{cases}$$

$RO\cdot$ radical can also induce the decomposition of ROOH:

$$\sim\sim CH_2-\underset{\underset{O\cdot}{|}}{\overset{\overset{CH_3}{|}}{C}}-CH_2\sim\sim + \sim\sim CH_2-\underset{\underset{OOH}{|}}{\overset{\overset{CH_3}{|}}{C}}-CH_2\sim\sim \longrightarrow \sim\sim CH_2-\underset{\underset{OH}{|}}{\overset{\overset{CH_3}{|}}{C}}-CH_2\sim\sim + \sim\sim CH_2-\underset{\underset{O}{\underset{\|}{O\cdot}}}{\overset{\overset{CH_3}{|}}{C}}-CH_2\sim\sim$$

Other reactions may also happen, such as the ROOH will first decompose to generate water and a biradical intermediate, which will further transfer to ketone.

$$\sim\sim CH_2-\overset{\overset{CH_3}{|}}{CH}-\underset{\underset{OOH}{|}}{CH}-\overset{\overset{CH_3}{|}}{CH}-CH_2\sim\sim \longrightarrow \left[CH_2-\overset{\overset{CH_3}{|}}{CH}-\underset{\underset{O\cdot}{|}}{C}-\overset{\overset{CH_3}{|}}{CH}-CH_2\right]_n + H_2O$$

$$\left[CH_2-\overset{\overset{CH_3}{|}}{CH}-\underset{\underset{O\cdot}{|}}{C}-\overset{\overset{CH_3}{|}}{CH}-CH_2\right]_n \longrightarrow \sim\sim CH_2-\overset{\overset{CH_3}{|}}{CH}-\overset{\overset{O}{\|}}{C}-\overset{\overset{CH_3}{|}}{CH}-CH_2\sim\sim$$

3. Carbonyl groups

Carbonyl groups can be generated from the above reaction, as well as the reaction of monomer and CO or the exposure of polymer to ozone. The following photo process is possible for carbonyl groups:

$$(>C=O) \xrightarrow{h\nu} {}^1(>C=O) \longrightarrow {}^3(>C=O)$$

Norrish explored the initiation mechanism of carbonyl groups, there are three different kinds of Norrish reactions:

(1) Norrish I reaction (α-position). Norrish I reaction will happen between carbonyl group and nearby α-position to form corresponding radicals:

$$R-\overset{O}{\overset{\|}{C}}-R_1 \xrightarrow{h\nu} \begin{cases} R-\overset{O}{\overset{\|}{C}}\cdot + \cdot R_1 \to R\cdot + C-O + R_1^* \\ R\cdot + R_1-\overset{O}{\overset{\|}{C}}\cdot \to R\cdot + C-O + R_1^* \end{cases}$$

(2) Norrish Ⅱ reaction. Norrish Ⅱ reaction is only happen at the γ-position of ketone, which is possible at room temperature. Normally, Norrish Ⅱ reaction is easier than Norrish Ⅰ reaction.

$$R-\overset{O}{\overset{\|}{C}}-\overset{\alpha}{CH_2}-\overset{\beta}{CH_2}-\overset{\gamma}{CH_2}-R_1 \xrightarrow{h\nu} R-\overset{O}{\overset{\|}{C}}-CH_3 + CH_2=CH-R_1$$

Aromatic ketone can also undergo Norrish Ⅱ reaction, such as the reaction of phenyl n-heptyl ketone.

$$CH_3-CH_2-CH_2-\overset{\gamma}{CH_2}-\overset{\beta}{CH_2}\vdots\overset{\alpha}{CH_2}-\overset{O}{\overset{\|}{C}}-\phi \xrightarrow{h\nu} CH_3-CH_2-CH_2-CH=CH_2 + CH_3-\overset{O}{\overset{\|}{C}}-\phi$$

The process of Norrish Ⅱ reaction can be explained according to the biradical intermediate mechanism. In the case of polymer, Norrish Ⅱ reaction happen at the middle of the chain.

$$R-\overset{O}{\overset{\|}{C}}-CH_2-CH_2-CH_2-R_1 \xrightarrow{h\nu} R-\overset{O-H}{\underset{CH_2-CH_2}{\overset{\diagup}{C}\diagdown}} \overset{\diagup}{\underset{}{CH-R_1}} \to$$

$$R-\overset{OH}{\underset{CH_2-CH_2}{\overset{|}{C}\cdot}}\overset{\diagup}{\underset{}{\cdot CH-R_1}} \to R-\overset{OH}{\underset{}{C}}=CH_2 + CH_2=CH-R_1$$
Enol   Alkene

↓ Cyclization   ↓

$$R-\overset{OH}{\underset{CH_2-CH_2}{\overset{|}{C}}}-\overset{\diagup}{\underset{}{CH-R_1}}$$ $$R-\overset{O}{\overset{\|}{C}}-CH_3$$
Cyclobutanol   Ketone

(3) Norrish Ⅲ reaction. In the process of Norrish Ⅲ reaction, β-H transfer and the cleavage of the C—C bond close to carbonyl group will give an aldehyde and an alkene.

$$R-\overset{O}{\overset{\|}{C}}-\underset{\alpha}{\overset{\overset{\beta}{CH_3}}{CH}}-R_1 \xrightarrow{h\nu} R-\overset{O}{\overset{\|}{C}}H + CH_2=CH-R_1$$

A second large class of polymers contains phenyl moieties, either in the backbone [polyphenylene oxide(PPO), PC, PET, polysulfone(PSU)] or as a side group (PS). Another common phenyl group photoreaction, when subjected to UV radiation, is the photo-Fries rearrangemen (see Fig. 3.7). Upon absorption of a photon, the n- or p-orbital of the chromophore is promoted to a singlet pexcited state. Bond scission occurs primarily at aromatic ether C—O bonds, and causes rearrangement or degradation of the polymer backbone. Because the photo-Fries rearrangement proceeds in a "caged" environment, it is independent of free volume and is almost independent of T_g.

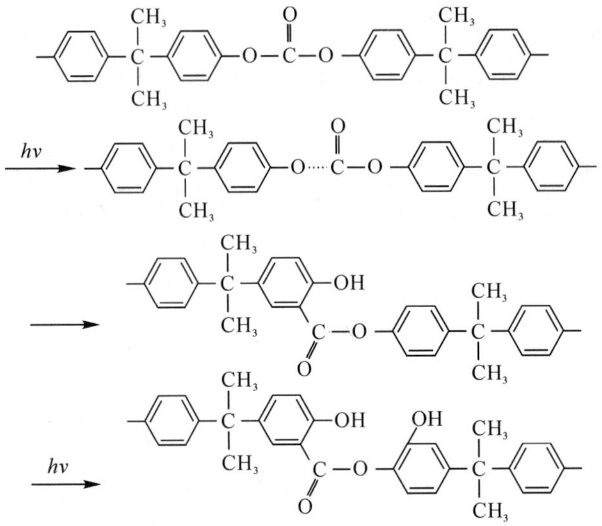

Fig.3.7 Photo-Fries rearrangement in bisphenol-A polycarbonate

The photo-Fries mechanism can be promoted with light in the region above 300 nm, and accounts for the yellowing of the polymer observed at long wavelengths. Chain scission, on the other hand, is promoted by light at shorter wavelengths. Because photo-Fries products are easily photodegraded, only small chain oxidation products (see Fig. 3.8).

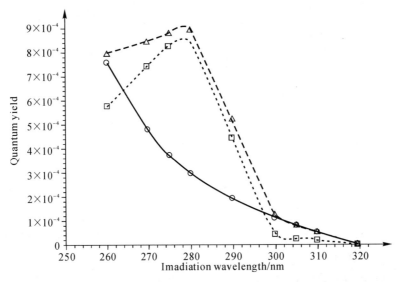

Fig.3.8 Quantum yield for chain scission (−o−), efficiency of photo-Fries rearrangement (−□−), and absorption spectrum (−−△−−) of bisphenol-A polycarbonate

Once photoradicals are formed, they can either recombine in the cage or diffuse to the outside for further reactions. The radicals formed can undergo a variety of abstraction and decomposition reactions, resulting in a multitude of chemical products. The propagation and recombination steps are similar to those encountered in thermal degradation, and will not be discussed further here. In the presence of UV > 240 nm and oxygen, the phenyl groups in PS, PPO, and PC can undergo ring-opening oxidation with formation of mucondialdehyde (see Fig. 3.9).

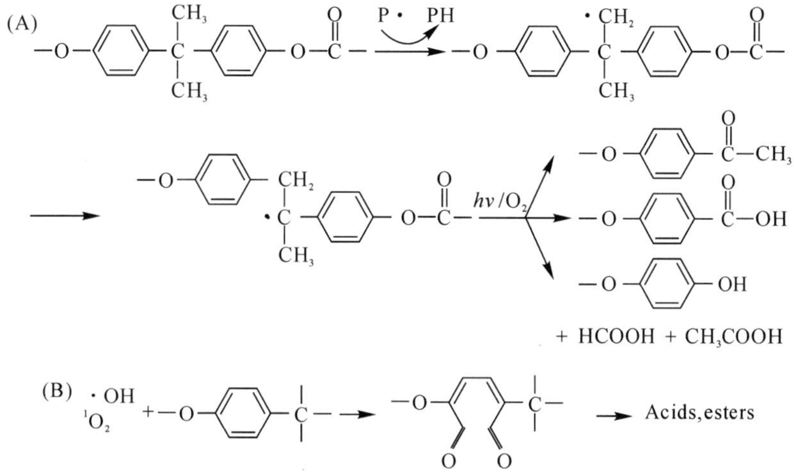

Fig. 3.9 Chain scission (A) and ring-opening (B) reactions in bisphenol-A polycarbonate

4. Singlet oxygen

$^1O_2$ is active species; $^3O_2$ is in ground state, it can be activated by light to form singlet $^1O_2$. $^1O_2$ is very active to react with unsaturated bond to form different kinds of ROOH, as mentioned by Carlson:

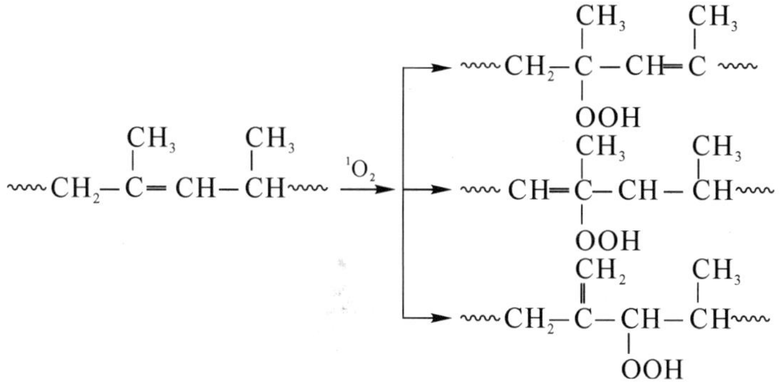

It should be emphasized that Dr. Mayo found singlet O_2 do not react with atactic polypropylene (a-PP).

3.3 Stabilization of photo degradation and photo-oxidative degradation

In principle, there are two kinds of methods to prevent polymers from photodegradation and photo-oxidative degradation: (1) change the structures of polymers to get more stable structures, such as purification of polymer, connection of stabilizers to the polymer structures during polymer synthesis or subsequent graft. However, this method is not very efficient due to the difficulty in practical applications. (2) addition of stabilizer. This is very simple and effective. There are four kinds of photo stabilizers: light shielding agent (UV screener), UV absorber, quenching agent and hindered amine light stabilizer (HALS).

1. Light shielding agent (UV screener)

UV screener can reduce the interaction between light and polymer to protect polymeric products from photo degradation. UV screener can function in the following way: reflection/absorption from the surface of polymer, and transmission to the inside of polymer. Basically, UV screener is a combination of physical and chemical process.

(1) Surface coating. Coating the polymer surface with metal film and paint is general method for surface coating. For example, paint is used to protect the wood products; coating of ABS products with metal film can protect them from photodegradation.

(2) Pigment. Addition of pigments to polymer is a useful method to improve the photo stability of polymer. The common pigments used in application are carbon black, TiO_2, ZnO, etc. Carbon black is the most effective antioxidant with good heat resistance. Due to its dark color, carbon black can only be used in dark-colored polymeric product, such as tire.

2. UV absorber

UV absorber can absorb UV light strongly, then, transform UV to heat or light with lower energy.

UV absorber is an important kind of photo stabilizer, including o-hydroxybenzophenone, phenyl salicylate, benzotriazole and hydroxybenzotriazine.

(1) o-hydroxybenzophenone. o-hydroxybenzophenone is an important kind of UV absorber, its structure is as following:

In such kind of structure, there must be a hydroxyl group on the *ortho*-position the carbonyl group. There is hydrogen bonding between hydroxyl groups. After absorption of photo, the hydrogen bonding is broken, the ketone structure is changed to enol form. After transform the absorbed photo energy to heat, the structure of the o-hydroxybenzophenone is regenerated. The more stable the hydrogen bonding, the more energy needed to cleave the ring; the less energy transfer to the polymer, the more stable. It is proved that the more stable the intramolecular H-bond, the easier the transformation of the ketone to enol form.

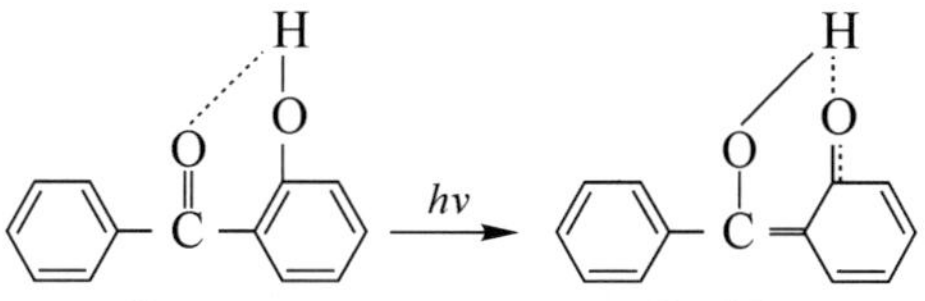

Chapter 3 Photo degradation and stabilization

[Ketone → Enol form scheme]

It should be mentioned that if there is no hydroxyl group on the *ortho*-position of the carbonyl group, the corresponding structure can also absorb UV light. However, the absorption will induce the decomposition of the structure, the decomposed product will work as a sensitizer to accelerate the degradation process. It can be explained as follow: the highly conjugated structure between ketone and the two phenyl ring in dibenzyl ketone make the transition of $n \to \pi$ is much easier. After the excitation of light, it will be excited to singlet state (n, π). Such excited molecules will undergo Norrish I reaction once it gain enough energy. The formed radicals will accelerate the decomposition of polymers.

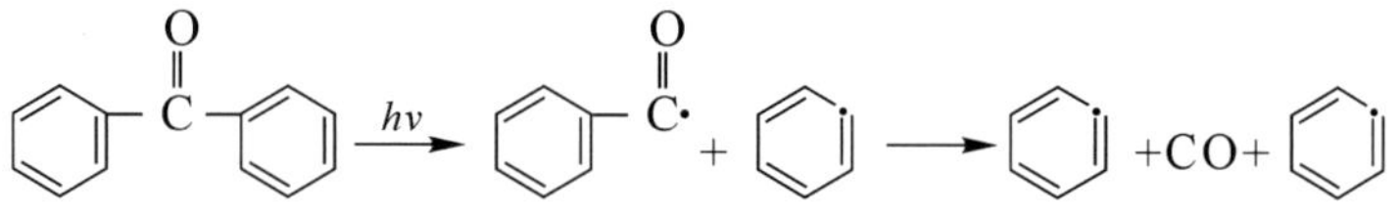

[Norrish I reaction scheme: dibenzyl ketone → benzoyl radical + phenyl radical → benzene + CO + phenyl radical]

(2) Phenyl salicylate.

[Structure of phenyl salicylate with OH on ortho position]

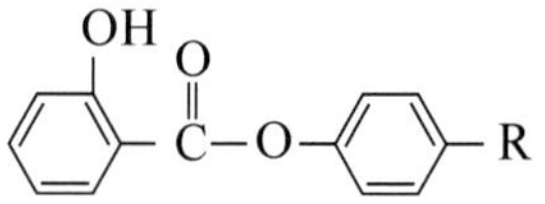

Phenyl salicylate is the earliest UV absorber, with advantage of cheap and good miscibility with polymers. However, it shows lower UV absorption efficiency and unstability to UV light. It will undergo photo-Fries rearrangement to form o-hydroxyl biphenyl ketone-like structure, which is explained above.

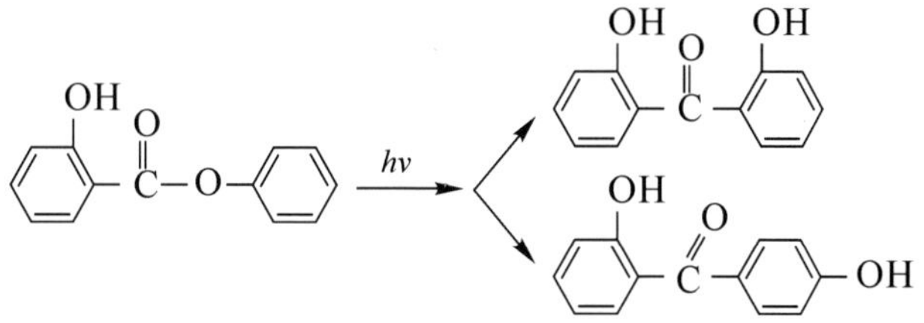

[Photo-Fries rearrangement scheme of phenyl salicylate]

Resorcinol benzoate is similar to phenyl salicylate, it can also undergo photo-Fries rearrangement:

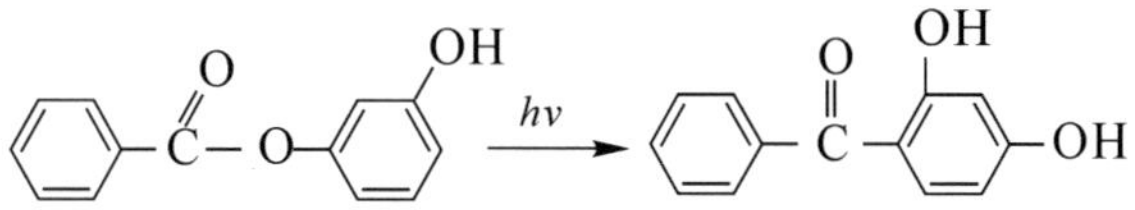

[Photo-Fries rearrangement scheme of resorcinol benzoate]

(3) Benzotriazole.

[Structure of benzotriazole UV absorber]

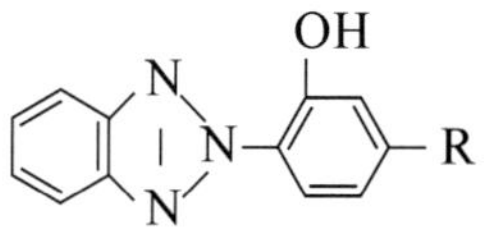

The hydroxyl groups and the triazole can form hydrogen bonding with each other. The absorption of photo can open the hydrogen bonding ring and convert the photo energy to non-radioactive heat, then go back to its original structure.

Benzotriazole can strongly absorb UV light from 280 nm to 380 nm and almost no absorption in the visible region, it also displays good thermal stability and less volatile. Due to the above advantages as well as no coloring to the polymeric products, it is broadly used in polyolefin and related products.

(4) Hydroxybenzotriazine.

Hydroxybenzotriazine was developed in 1970s. It can strongly absorb UV light from 300 nm to 400 nm, which is more efficient than benzotriazole. It is proved that the more hydroxyl groups in the structure, the more efficient the UV absorber. For example, 3,3′,3″-triphenyl-3,3′,3″-[1,3,5] triazinane-1,3,5-triyl-tris-propionic acid trimethyl ester is a powerful UV absorber with 3 hydroxyl groups.

3. Quenching agent

The mechanism of quenching agent is different from UV absorber. It neither absorbs UV light strongly nor transfers the photo energy to heat through intermolecular structure transformation. Quenching agent can undergo intermolecular energy transfer, then, quench the energy of excited state, and return to ground state. Therefore, the most popular quenching agent is transition metals, such as Ni and Cu.

Generally, there are two pathways for the quenching agent.

(1) The excited polymer transfer its energy to quenching agent, making it a non-reactive excited molecule and deplete the energy through photophysics process:

$$R + Q \longrightarrow R + Q \rightsquigarrow Q$$

Excited polymer   Quenching agent

Chapter 3 Photo degradation and stabilization

(2) The excited polymer forms excimer with quenching agent, and lose energy through the photophysics process of this excimer:

$$R + Q \longrightarrow [R \cdots\cdots Q] \rightsquigarrow R + Q$$

4. Hindered amine light stabilizer (HALS)

In 1962, M. B. Neiman from former Soviet Union reported that hindered nitroxide is very effective for the photo stability of polymers. In 1973, the first mine light stabilizer is developed in Japan with commercial name of Sanol. LS-744.

R_1, R_2: any groups
R_3: $-H, -CH_3, -C_2H_5$

HALS is a multiple functioned light stabilizer, it can capture radical, decompose ROOH, energy transfer of the excited state and capture singlet oxygen.

(1) Capture of radical. HALS can form nitroxide radical in the presence of photo and oxygen.

$$\rangle N-H \xrightarrow[\text{[O]}]{hv} \rangle N-O\cdot$$

The nitroxide radical can capture free radical very efficiently:

$$\rangle N-O\cdot + R\cdot \longrightarrow \rangle N-O-R$$

$$\rangle N-O\cdot + ROO\cdot \longrightarrow \rangle N-O-R + O_2$$

The nitroxide radical can be regenerated and capture radical again:

$$\rangle N-O-R + ROO\cdot \longrightarrow \rangle N-O\cdot + R_1OOR$$

(2) Quenching of excited state. HALS can quench excited molecules including singlet oxygen, which is similar to Ni complex. It is demonstrated that the quenching ability of HALS (amine) before oxidization is very weak. After oxidation to nitroxide, it can efficiently quench the excited states, even stronger than Ni complex.

(3) Decomposition of ROOH. HALS can decompose ROOH and destroy its accumulation, which can inhibit the automatic oxidation process. During the decomposition of ROOH by HALS, nitroxide radical is generated and can further capture radicals:

$$\rangle N-H + ROOH \longrightarrow \rangle N-OH + ROH$$

$$2 \rangle N-OH \longrightarrow 2 \rangle N-O\cdot + H_2$$

Due to its advantages, HALS becomes a very important kind of light stabilizer, more and more HALS have been developed in the following direction: 1) higher molecular weight HALS (2,000 - 3,000) to overcome its fastpenetration; 2) non-basic or less basic HALS to overcome its strong basic property; 3) HALS with non-piperidine structures, such as piperazine-containing HALS; 4) more functions, such as introduction of UV absorbs and antioxidant structures to HALS; 5) reactive HALS, which can grafted to polymeric structures during the processing to form long term stable polymeric products.

3.4 Photodegradable polymers

Besides photo stabilization, there is another research topic "photodegradable polymer". If a polymer can be decomposed through photo degradation, the decomposed polymer can further decomposed by microorganisms and give a "green cycling" of polymer. Therefore, photodegradable polymer becomes a hot topic.

Generally, there are some research topics for photodegradable polymer.

(1) Change the polymer structures. The presence of branched chains, double bonds, carbonyl groups can accelerate the photodegradation of polymers. If functional groups that are sensitive to UV light can be introduced into the polymer structures, the polymer can be decomposed upon photo-oxidative degradation. There are several methods to introduce such kind of groups: ① irradiation of PE in the presence of oxygen or CO, carbonyl groups can be introduced to PE structure. ② monomers with carbonyl groups can be introduced to PE upon the irradiation. ③ photo sensitive groups can be introduced to certain kind of polymer through so called "ecolyte method". For example, the copolymerization of vinyl ketone or CO with vinyl monomers can give photodegradable PE.

(2) Photodegradable polymer through photo sensitizers. Photo degradation is a radical process, thus, photosensitive groups can be introduced to polymeric structures in order to achieve photo degradation. Upon the irradiation, large number of radicals is generated to achieve fast photodegradation.

(3) Photodegradable masterbatches. In order to keep the lifetime of polymeric products in a certain period, photodegradable masterbatches are developed. For example, during the processing of polyolefin, a certain percentage of photodegradable masterbatches is added to achieve photo stabilities in certain period and photodegradable properties.

Questions

1. Describe two methods used to prepare photodegradable polymeric materials.
2. Hindered amine light stabilizer (HALS) is a very useful photo stabilizer. Write the reactions associated with the following HALS.
3. For photo-oxidative degradation, there are several Norrish reactions. Explain the photo-

oxidative degradation of the following compound and write the possible mechanism.

Further reading Ⅰ: photo degradation of polymers

Polymers will change over time when exposed to UV radiation. These changes are the result of light-induced homolytic fission of chemical bonds (photolysis) and photo-oxidation.

The UV resistance of unprotected polymers varies widely and depends on the structure and composition. In general, polymers with high bond energies and without UV sensitive groups are very stable, whereas those with weak bonds and high concentration of chromophoric groups are very sensitive to UV degradation. Another important factor is the accessibility of reactive hydrogen on secondary and tertiary carbon atoms. In the case of polyolefins, the UV stability decreases in the order:

$$\left[\begin{array}{c}HC-CH_2\\|\\CH\\/\backslash\\H_3C\ CH_3\end{array}\right]_n > \left[\begin{array}{c}HC-CH_2\\|\\H_3C\end{array}\right]_n > \left[CH_2-CH_2\right]_n > \left[CH_2-CH_2\right]_n$$
$$\text{Branched} \qquad \text{Linear}$$

The UV stability will also depend on the concentration and type of impurities like residual catalyst and thermal degradation by-products from melt processing of the resin. These contaminations can either directly initiate photo degradation or can sensitize the polymer. The addition of pigments and fillers will usually improve the UV stability. For example, in the case of carbon black, the degradation rate of black PE films vary directly with particle size and inversely with concentration.

Typical initiators for photo degradation are ketones, quinones and peroxides because they absorb light below 400 nm which causes bond excitation and cleavage to radicals. Thus, photo degradation may occur in the absence of oxygen but will be greatly accelerated by oxygen.

Photo degradation usually starts at the surface with visible cracks and discoloration, which leads to rapid loss of mechanical properties. In theory, olefins such as polyethylene (PE) and polypropylene (PP) are stable when exposed to sunlight, because both C—C and C—H bonds do not absorb UV radiation from the sun. However, there are often contaminations present which can initiate photo degradation resulting in surface cracking and loss of mechanical properties.

There are several mechanisms that can contribute to photo degradation. For example, if free radicals are directly produced by UV radiation, then all subsequent reactions are similar to those of thermal degradation, including chain scission, crosslinking, and secondary oxidation. Another mechanism of photo oxidation includes the photochemical production of electronically excited oxygen. In this case, degradation takes place by direct oxidation, for example, by reaction with allylic hydrogen which then could undergo chain scission with formation of terminal ketone groups:

$$\text{wwC=C-C ww} \xrightarrow{^1O_2} \text{wwC=C-C ww} \longrightarrow$$

(with H on middle C, then HOO on middle C)

$$\text{wwC=C-C ww} \xrightarrow{h\nu} \overset{O}{C}\text{ww} + \overset{\cdot}{C}=CH$$

(with HOO group)

Another contributor to photo degradation is ketone photolysis which proceeds via Norrish reaction. Ketones are formed by either thermolysis or photolysis of hydro peroxides. If these groups are exposed to light, they undergo chain scission via Norrish I or II reaction:

$$\text{wCH}_2-\text{CH}_2-\overset{O}{\underset{\|}{C}}-\text{CH}_2-\text{CH}_2\text{w} \xrightarrow{h\nu} \begin{cases} \text{I}: \text{wCH}_2-\text{CH}_2-\overset{O}{\underset{\|}{C}}-\text{C}\cdot + \text{H}_2\text{C}\cdot-\text{CH}_2\text{w} \\ \text{II}: \text{wCH}_2-\text{CH}_2-\overset{O}{\underset{\|}{C}}-\text{CH}_3 + \text{H}_2\text{C}=\text{CH w} \end{cases}$$

The Norrish I reaction leads to chain scission and formation of radicals that might then initiate photo-oxidation. However, in the case of polyethylene, homolytic cleavage processes (Norrish I) are only responsible for a small percentage of the chain breaks at room temperature.

Further reading II : hindered amine light stabilizers (HALS)

One of the most important classes of antioxidant for long-term heat protection of polymers are the so called hindered amine light stabilizers (HALS) which are very effective inhibitors against free radical induced degradation of polymers at low and medium temperatures. This class of amine stabilizers is based on 2, 2, 6, 6-tetramethyl-piperidine derivatives. They are often employed as light stabilizers, particularly for olefins which explains their naming.

HALS can be categorized according to their molecular weight: HALS with low molecular weight of about 200 to 500 are commonly referred as low MW HALS, while those with a molecular weight of 2,000 or higher are referred as high MW HALS. These two classes have different rates of diffusion in the polymer matrix, which is an important factor in protecting polymers from UV radiation. In general, high MW HALS are more effective long-term heat stabilizer than low MW HALS, whereas very low MW HALS do not provide much thermal stability at all. The efficiency of HALS can also be greatly affected by the structure, e. g. side groups, and sometimes by the formulation and surface conditions. To maximize the light stability of plastics, the most suitable HALS structure must be chosen.

Photo-oxidation usually starts at the surface with visible cracks, which leads to rapid degradation of the mechanical properties. In theory, olefins such as polyethylene (PE) and polypropylene (PP) are stable when exposed to sunlight, because both C—C and C—H bonds do not absorb UV radiation from the sun. However, there are always contaminations present like residual catalyst, pigments, and thermal degradation by-products from melt processing of the resins that can initiate photo degradation of any polyolefin when exposed to UV light, which, in turn, causes chain scission and/or crosslinking and eventually surface cracking and loss of mechanical properties.

The high efficiency of HALS is based on a complex set of free-radical scavenging reactions, which starts with the oxidation of HALS to nitroxide radicals (RNO·) by reacting with peroxides or hydroperoxides, which subsequently react with polymeric alkyl radicals, yielding hydroxylamine ethers which can react with a variety of compounds. New RNO· radicals may be regenerated from the reaction of hydroxylamine ethers with alkylperoxy and acylperoxy radicals (Denisov cycle) as shown following:

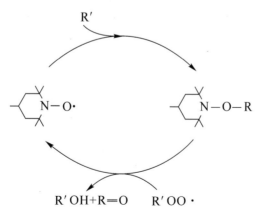

The oxidation of HALS to the NO· radicals is a relatively slow and temperature-dependent reaction. Furthermore, they are not very efficient at elevated temperatures (above 80 ℃ or so), thus, they are not very effective processing stabilizers. For these reasons, HALS are often used in combination with primary and secondary antioxidants. This combination may also show synergistic or antagonistic effects. Due to these unpredictable effects, complex mixtures are often used. For this reason, the study of interactions between stabilizers from different chemical classes is of great importance.

Further reading Ⅲ : metal deactivators and UV-light absorbers

Metals can accelerate thermal oxidation of polyolefins and related plastics such as EPDM. To prevent metal catalyzed degradation, it is often neccessary to combine an antioxidant with a metal deactivator if the former does not contain a moity with radical scavenging properties.

The function of a metal deactivator or metal-deactivating agent (MDA) is to form an inactive complex with the catalytically active metal ion. Most chelating agents are well suited for this purpose because they form thermally stable metal complexes. The general feature of chelating agents is their polyfunctionalities. They contain several ligand atoms such as N, O, S, P often in combination with functional groups such as hydroxyl, carbonyl, or carbamide. The different substituents will affect the properties, like polarity, volatility, miscibility, and UV absorption spectrum. Common metal deactivators include salicylidene propylenediamine, mercaptobenzothiazoles, mercaptobenzimidazoles and many thiadiazole and triazole derivatives.

Many metal deactivators also function as UV-absorbers by competitive absorption of destructive UV radiation. These compounds generally absorb UV light much more strongly than the polymers that they protect. The excited states of these compounds relax to the ground state very rapidly and efficiently without producing any radiation, which imparts high UV stability.

Benzophenone   Benzotriazole

UV-absorbers are usually only effective in thick and unpigmented/unfilled plastic products, especially when combined with HALS. They are usually not recommended for thin plastic films (below 100 μm) because a certain thickness is required to achieve sufficient UV-absorption.

Another important class of light stabilizers are nickel quenchers. These compounds "quench" the excited states of carbonyl groups, that form during photo oxidation and through decomposition of hydroperoxides. However, these compounds are not widely used because they contain heavy metal and cause discoloration.

Further reading Ⅳ: free radical photoinitiators

Photoinitiators are compounds that produce radicals when exposed to UV light. These then react with monomers and/or oligomers to initiate polymer chain growth. They are essential ingredients of all UV-curable adhesives, inks and coatings.

Free radical UV photoinitiators can be classified into Norrish type Ⅰ and Ⅱ initiators.

1. Norrish type Ⅰ

Initiators are typically compounds containing benzoyl groups. The carbonyl group of the initiator absorbs a photon and is transformed into an excited state. Subsequent homolytic cleavage of the excited α-carbon bond produces two radical fragments. For instance, the cleavage of 2,2-dimethoxy-1,2-diphenyl-ethan-1-one largely yields a methoxybenzyl and benzoyl radical. The benzoyl radical initiate free radical polymerization whereas the methoxybenzyl radical decomposes to give the more stable methyl radical and methyl benzoate.

Two other very common Norrish type I initiators are 1-hydroxycyclohexylphenyl-ketone and 2-hydroxy-2-methyl-1-phenylpropanone. Both are soluble in many solvents (monomers) and can be combined with other initiators.

2. Norrish type Ⅱ

Initiators absorb UV light to form excited molecules which then abstract an electron or hydrogen atom from a donor molecule (synergist). The donor molecule then reacts with a monomer to initiate polymerization. Common type II photo initiator systems include benzophenone and its derivative and isopropyl thioxanthone in combination with a synergist such as tertiary amines. The amines function as active hydrogen donors for the excited photoinitiators. The abstraction of hydrogen produces very reactive alkyl-amino radicals which subsequently initiate polymerization.

Tertiary amines also reduce air inhibition; oxygen, which diffuses into the mixture, reacts with growing free-radical polymer chains to form unreactive peroxy radicals. The tertiary amines react with these radicals converting them to reactive alkyl-amino radicals, and thereby prevent air inhibition.

$$R\cdot + O_2 \longrightarrow ROO\cdot + R-N(R)(R)$$

Peroxyl radical   Tertiary amine

$$\longrightarrow ROOH + R-\dot{N}(R)(R)$$

Organic hydroperoxide   Tertiary amino radical

Two frequently used aromatic amine synergists are 2-ethylhexyl-(4-N, N-dimethyl amino)benzoate, and 2-ethyl-(4-N, N-dimethylamino)benzoate.

Another important class of amine synergists is acrylated amines. The acrylate on the amine synergist will allow it to react with the chain radicals and thereby reduce the potential for surface migration (less tack). They also provide faster surface cure, reduced volatility (odor), and increased solvent resistance.

Table 3.2 shows the degradation of different kinds of ketones.

Table 3.2 **Degradation of different kinds of ketones**

Compound	Chemical structure	Type
2,2-Dimethoxy-1,2-diphenylethan-1-one	Ph-C(OCH$_3$)(OCH$_3$)-C(=O)-Ph	α-Cleavage
2-Hydroxy-2-methyl-1-phenylpropanone	HO-C(OCH$_3$)(OCH$_3$)-C(=O)-Ph	α-Cleavage
1-Hydroxy-cyclohexylphenylketone	(cyclohexyl-OH)-C(=O)-Ph	α-Cleavage
Benzophenone	Ph-C(=O)-Ph	Hydrogen abstraction

Continued table

Compound	Chemical structure	Type
Isopropyl thioxanthone	(structure)	Hydrogen Abstraction
2-Ethylhexyl-(4-N,N-dimethyl amino)benzoate	(structure)	Synergist
Ethyl-4-(dimethylamino) benzoate	(structure)	Synergist

Further reading V: radiolytic degradation

1. Interaction of high-energy radiation with matter

Understanding the effects of ionizing radiation on the properties of plastics materials is important in nuclear engineering, space research, radiation processing, and radiation sterilization.

Radiation composed of photons or particles with energy much higher than those encountered in electron bonding orbitals is referred as high-energy radiation (X- or γ-rays, high-energy electrons, protons, and charged particles). Owing to this excess energy, high-energy radiation can penetrate much deeper into material to create ions, superexcited states, and hot radicals. The degradative effects are also much more extensive than with UV. The principal sources of high-energy radiation are electron beam accelerators, which account for 90% of commercial radiation capacity; the remainder consist of ^{60}Co installations. ^{60}Co is unstable and decays to the stable ^{60}Ni according to following equation:

$$^{60}Co(t_{1/2}=5.27y) = {}^{60}Ni* + \beta^-$$

The nuclei of Ni atoms that result from this decay are in an excited state and immediately emit two γ-rays of energies 1.332 MeV and 1.173 MeV. The low-energy is bare absorbed by the ^{60}Co housing and all the radiolytic effects result from the γ-ray emission.

Electron accelerators can deliver higher dose rates, whereas ^{60}Co sources are characterized by a greater depth of penetration.

A fast electron loses most of its kinetic energy by inelastic collisions with electrons from the medium, producing energetic secondary electrons. Depending on the energy of the radiation, many secondary electrons of decreasing energy will be created (δ-rays). Owing to the augmentation of linear energy transfer (LET) with decreasing kinetic energy, the fast electrons lost most of their kinetic energy near the end of the track, resulting in local regions of highly reactive species called "short tracks", "blobs", and "spurs" (see Fig. 3.10).

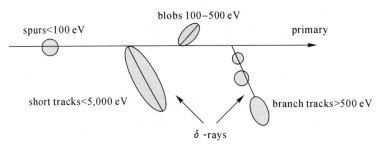

Fig. 3.10 Schematic representation of a fast electron track

In ^{60}Co irradiation, the primary interaction is Compton scattering between the high-energy photon and an atomic electron. Only part of the photon energy is transferred to the electron and the resultant photon of lower energy is scattered. The process will continue with creation of more scattered electrons and secondary electrons (see Fig. 3.11). Except for the primary event, the effects of g-rays and accelerated electron irradiation are similar. From a single incoming γ-photon or highenergy electron, a shower of secondary electrons is generated which is responsible for most of the ensuing chemical reactions.

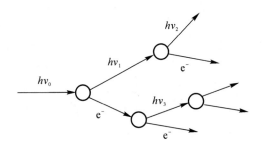

Fig. 3.11 Compton scattering and fate of the secondary electrons

2. Radiation chemistry

Radiation chemistry is now a well-established area of science that deals with chemical and physicochemical changes produced by the absorption of high-energy radiation (HER) by matter. The subject covers events that occur from the passage of the ionizing particle to the completion of chemical reactions. To encompass the enormous time span for the various processes, it is usual to distinguish between the "physical" stage ($10^{-17} - 10^{-11}$ s) when energy from the incident particle is eposited into localized regions of space ("spurr"), the "physicochemical" stage ($10^{-12} - 10^{-3}$ s) during which reactive species (ions, quasi-free electrons, excited molecules, hot radicals) are formed and react within the "spurs", and the "chemical" stage ($10^{-9} - 10^{-3}$ s in solution, several days in solid polymers) where stable

Chapter 3 Photo degradation and stabilization

species (radicals, trapped electrons, cations) diffuse and react outside the clusters. The standard unit of absorbed dose is the Gray (Gy), defined as the energy imparted by the high-energy radiation to a mass of matter equivalent to 1 J/kg (1 Gray = 100 rad). Molecular changes are characterized by a G factor, in units of μ mol/J, defined as the event yield per 100 eV of absorbed energy.

Although much discussion has occurred in the past on the relative importance of radical and ionic reactions, it is now established that the major chemical changes in irradiated polymers are accounted for by free-radical reactions. In the early stage of the reactions, the reactive species are concentrated in "spurs" and particle "tracks" in a manner similar to their parent ionized or excited molecules. The kinetics at this point must take into account the inhomogeneous distribution of the radicals, before they can diffuse away. Ionizing radiation is unique in the sense that reactions can be initiated randomly at any temperature. Cryogenic temperatures, at 4 K and below for instance, have been used extensively to prolong the macroradicals' lifetime for electron spin resonance (ESR) measurements. Apart from the mode of initiation, all other material changes brought about by high-energy radiation are governed by radical reactions, in perfect analogy to those generated by other means, such as photochemical, thermal, or mechanochemical degradation. As a matter of fact, despite the enormous difference in HER energy (106 eV) and molecular binding energy ($\sim$5 eV), the chemical effects of HER can best be compared with those of UV light with energy in the 5 - 20 eV range.

The effects of ionizing radiation depend greatly on the structure of the polymer, the temperature, and the nature of environment. A 50% loss in ultimate elongation (a common measure of the effect of irradiation) (see Table 3.3), for instance, can vary from doses as low as 3.5 kGy for PTFE, to more than 4,000 kGy for PS, polyimide, or polyaryletheretherketone (PEEK) (see Table 3.3). The unusual radiation sensitivity of PTFE is attributed to the unique stability of perfluoro macroradicals which favors chain scission over crosslinking. In the presence of air, these fluorine-containing radicals are converted into peroxy radicals which degrade readily into low MW fragments. PTFE can be crosslinked by HER when irradiated in an inert atmosphere above its melting point (330-340 ℃). Polymers containing phenyl groups owe much of their radiation resistance to excited-state energy transfer to the benzene rings, which act as excited state quenchers. Although "energy transfer" is the widely accepted protection mechanism in the HER degradation of aromatic compounds, it does not explain why styrene does not show the same protective effect as polystyrene. One alternative suggestion is that H atoms resulting from the primary effect of radiation are added to the aromatic rings, and are no longer able to produce secondary macroradicals by abstraction.

Table 3.3 Relative radiation stability in air of major commercial polymers (based on a 50% decrease in ultimate elongation)

Polymer	Dose/kGy
Polytetrafuoroethylene	10
Polytrifluorocholore thylene	30
Polymethyl methagylate	300
Polycaprolactam	600
Isotactic polypropylene	1,000
High-density polyethylene	1,000
Polyvinyl chloride	1,500
Polyethylene terephthalate	2,000
Polytriethylene glycol dimethacrylate	2,000
Low-density polyethylene	3,000
Polyurethanes	4,000
Melaminoformaldehyde resin	5,000
Polycarbonates	5,000
Polystyrene	5,000
Epoxy resin ED-10	15,000
Epoxy resin ETZ-10	30,000
Polyimides	100,000

The two fundamental processes that result from radiochemical reactions are chain scission and crosslinking, characterized by G_s and G_x, respectively. If $G_s < 4G_x$, branched polymers can be formed and may eventually evolve into a three-dimensional network structure. Based on the assumptions that: the initial polymer molecular weight distribution (MWD) follows a random distribution, cross-linking occurs by H-linking, and crosslinking and scission occur with random spatial distribution (without clustering).

Charlesby and Pinner have shown that the sol fraction(s) should follow the following equation, where R [kGy] is the absorbed dose.

$$s + s^{0.5} = 0.96 \times 10^5 / (R \cdot M_w \cdot G_X) + (G_S / 2G_X)$$

The Charlesby-Pinner plot of $(s + s^{0.5})$ as a function of $1/R$ gives a straight line, with slope $\propto 1/G_x$, and an intercept on the ordinate equal to $G_s/2G_x$. Departure of the plot from a straight line may originate from deviation from a random distribution, or from chemical heterogeneities in copolymers. Regardless of the shape of the curve, the incipient dose for gelation is determined, by definition, from the sol curve at $s + s^{0.5} = 2$ (see Fig. 3.12).

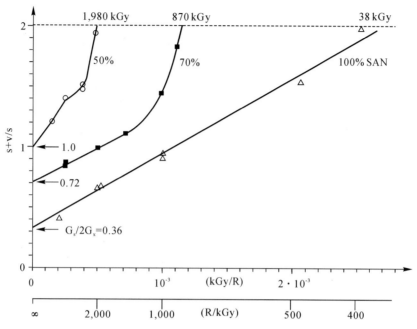

Fig. 3.12   Charlesby-Pinner plot of irradiated compatible PMMA-SAN blends

The sol and gel fractions refer solely to the SAN component in the mixture and are corrected for grafting to PMMA. Large curvatures of the plot for blends are interpreted as a result of chemical interactions between the two components.

3. Radiolysis stabilization

It has been known since the early days of radiation chemistry that some simple organic compounds, such as benzene, halogenated hydrocarbons, nitriles, amines, and alcohols, can protect the polymers from the deleterious effects of high-energy radiation. Many of these "antirad" substances interfere at some stage with the radiolytic degradation scheme, as depicted in Table 3.4, to reduce damage to the plastics.

The most efficient present-day "antirad" agents are antioxidants which act essentially in the chemical stage by scavenging free radicals in a similar way to that in the other types of degradation (see Table 3.4). Aromatic compounds are highly efficient at quenching excess energy of excited states formed by geminal recombination. Because most commercial antioxidants have aromatic rings in their structures, they can also act as primary absorbers by diverting the radiation energy into harmless vibrational energy, as in following equations:

$$P \xrightarrow{\gamma_1 e^-} \{P^+ + e^-\}_{spur} \longrightarrow P* \longrightarrow 2R*$$
$$P* + Q \longrightarrow P + Q*$$

Table 3.4 Polymer protection scheme during radiolysis

Radiolysis stage	Physical or chemical event	Protection means	Protection mechanism
Physical physicochemical	Energy absorption molecular ionization	Sheets of lead or concrete positive ion scavengers	Decrease in radiation intensity transfer of one electron to polymer cation without subsequent excitation
Chemical	Geminate recombination	Electron scavengers	Acceptance of ejected electron and lower probability of dissociative recombination
	Dissociative states	Dledtronic excited-state quenchers	Divergence of excitation energy into heat or longer-wavelength electromagnetic radiation
	Cleavage of C—H bonds	H atom donors	Transfer of H atoms to macroradicals (reparation mechanism)
	Diffusion and abstraction reactions of mobile H atoms	H atom acceptors	Acceptance of H atoms, preventing formation of more radicals
	Reactions of macroradicals	Radical scavengers	Reaction with macroradicals, formation of stable species
	Reactions of macroradicals with oxygen benzop henone derivatives	Surtac coatings	Decrease in oxygen diffusion
		Oxygen absorbers	React with molecular oxygen under irradiation
		Antioxidants	Disrupt oxidation chain reaction by converting peroxy radicals into stable products
	Chain reaction of peroxy radicals (radio-oxidation)	Peroxide decomposers	Catalytic destruction of peraxy radicals

Aromatic polymers owe much of their radiation resistance to this "internal protection" effect. An efficient method of radiation protection would be to blend the degradative polymer with a radiation-resistant one. Energy transfer (Förster type) is efficient only at short range and the effect is most noticeable with compatible blends, such as PMMA and SAN.

4. Applications

High-energy radiation can penetrate deeply into organic materials, and can initiate

Chapter 3 Photo degradation and stabilization

reactions at low temperature without added chemicals or catalysts. These unique features are exploited in an increasing number of industrial applications, particularly in the biomedical field, in which chemical contamination or thermal degradation should be avoided. Radiation sterilization of medical commodities, one of the early achievements of radiation engineering, continues to increase its market share to the detriment of standard methods such as chemical sterilization with *di*-ethyl ether, or heat treatment. In the examples mentioned, the value of the finished products is generally high and the cost of radiation processing does not enter into consideration. In other applications, where the product is inexpensive radiation processing can still be economically viable if the quantity of radiation energy required is low. Radiation treatment of food and radiation-induced grafting or crosslinking of certain plastics belong to the latter category. Food is commonly irradiated at low doses in the 0.1 - 10 kGy range, in ordinary boxes or containers. Food preservation by irradiation is gaining acceptance for an ever-increasing number of agrochemical products such as spices, vegetables, and processed meat. This technique is currently viewed as the most effective of the available alternatives (cold storage, heat treatment, fumigants).

5. Radiation Sterilization

Disposable medical products are required by legislation to be sterilized prior to use. Among the four basic methods of sterilization-heat, ethylene oxide, gamma irradiation (^{60}Co), and electron beam techniques-the last has steadily increased its market share, largely at the expense of the diethyl ether method as a result of regulatory concerns about the quantity of chemical residuals. The Federal Drug Administration (FDA) in the US allows a device to be labeled as sterile if less than one device out of a batch of one million can show biological contamination. The standard dose to achieve this level is 25 kGy. Medical plastics that need to be sterilized must withstand at least this dose. The data given in Table 3.3 can be used as an initial estimate in material selection. Materials which are likely to degrade, such as PTFE or POM, should be avoided. Polymers containing aromatic groups have much greater radiation resistance than those with an aliphatic structure. Similar improvements can be obtained with crosslinks, and most thermosets and elastomers can withstand at least one radiation sterilization. The use of antioxidant additives can significantly offset the effects of radiation, and even a radiation-sensitive polymer such as polypropylene can withstand several radiation sterilization cycles when properly stabilized.

6. Controlled degradation and erosslinking

The reactions of chain scission, crosslinking, and grafting initiated by ionizing radiation have found many important applications in the plastics and rubber industries. Although high-energy radiation is destructive for most materials, this propensity can be usefully exploited in the controlled degradation of several polymers, such as PTFE, PEO, PP and cellulose. The most well-known example of degradation application is the manufacture of PTFE powders. Undegraded PTFE is too tough and slippery to grind. After submitting the polymer to a high dose range (500 – 1,000 kGy), the degraded PTFE becomes brittle and can be converted to fine particles while conserving its low surface energy. The PTFE powder is used as a solid lubricant, or formulated as grease for high-temperature applications.

Radiation crosslinking is another way to tailor properties to specific applications. High-energy radiation provides a clean and cost-effective method to achieve this means. The technique has been applied in the crosslinking of high-purity medical products, such as gloves and condoms, orthopedic ultra-high MW PE hip joints, and biocompatible hydrogels. Other important industrial applications of radiation crosslinking include the production of heat-shrinkable products: wires, cables, and tubing.

Chapter 4   Ozone degradation and stabilization

There are trace amounts of ozone in air, which can induce ozone degradation of polymer due to its highly oxidative activity.

4.1   Ozone degradation

The concentration of ozone is not very high on earth. Oxygen molecules can absorb UV from 110 - 220 nm and form ozone at high altitude as high as 20-30 km. The concentration of ozone at this altitude can achieve 500×10^{-8}, which can absorb the UV light from sun and protect life on the earth. The ozone can diffuse to ground, which is a main source of ozone on the earth.

Meanwhile, ozone can form on earth starting from NO_2 through the following reactions:

$$NO_2 \xrightarrow{h\nu} NO + O$$
$$O_1 | O_2 \longrightarrow O_3$$

Ozone is not stable, it can slowly decompose to oxygen. The oxidation activity of ozone is stronger than oxygen, is a strong oxidant.

It can react with different polymers, such as polyolefin and rubber.

1. Reaction with polyolefin (radical process)

Ozone can oxidize polyolefin to a mixture, with 30%-40% with —COOH, 60%-70% with —CO groups. The polymeric ROO· can also react with PE:

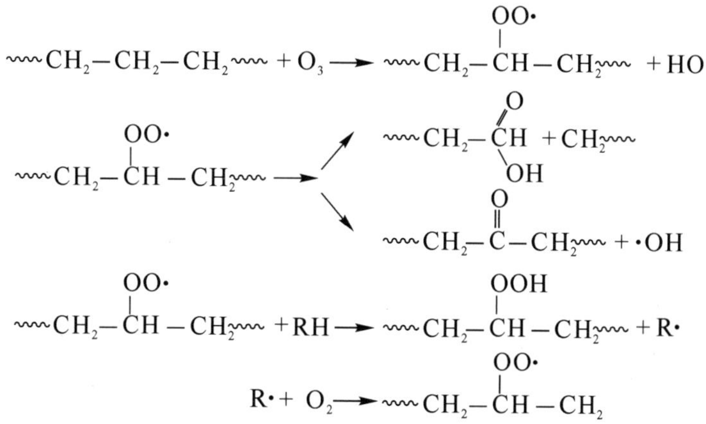

$$\sim\sim CH_2-CH_2-CH_2\sim\sim + O_3 \longrightarrow \sim\sim CH_2-\underset{\underset{OO\cdot}{|}}{CH}-CH_2\sim\sim + HO\cdot$$

$$\sim\sim CH_2-\underset{\underset{OO\cdot}{|}}{CH}-CH_2\sim\sim \longrightarrow \begin{array}{l} \sim\sim CH_2-\overset{\overset{O}{\diagup}}{CH} + CH_2\sim\sim \\ \diagdown OH \\ \sim\sim CH_2-\overset{\overset{O}{\|}}{C}-CH_2\sim\sim + \cdot OH \end{array}$$

$$\sim\sim CH_2-\underset{\underset{OO\cdot}{|}}{CH}-CH_2\sim\sim + RH \longrightarrow \sim\sim CH_2-\underset{\underset{OOH}{|}}{CH}-CH_2\sim\sim + R\cdot$$

$$R\cdot + O_2 \longrightarrow \sim\sim CH_2-\underset{\underset{OO\cdot}{|}}{CH}-CH_2$$

2. Reaction with PS

Ozone can also react with PS as following, which make the PS brittle and coloring:

$$\text{wwwCH}_2\text{-CH}\text{www} + O_3 \longrightarrow \begin{cases} \text{epoxide/ozonide product} \\ \text{wwwCH}_2\text{-}\overset{OO\cdot}{\underset{|}{C}}\text{H}\text{www (with phenyl)} \end{cases}$$

3. Reaction with Nylon-66

As shown in the following, reaction between ozone and Nylon-66 can generate HOO· raidcals, which can further induce polymer degradation:

$$-\overset{O}{\underset{\|}{C}}-\overset{H}{\underset{|}{N}}- + O_3 \longrightarrow -\overset{O}{\underset{\|}{C}}-\overset{OOOH}{\underset{|}{N}}- \longrightarrow -\overset{O}{\underset{\|}{C}}-\overset{O\cdot}{\underset{|}{N}}- + HOO\cdot$$

4. Reaction with rubber

The reaction between ozone and natural rubber is called ozone cracking, especially in presence of external stress, which can induce the cracking of rubber and corresponding products. The reaction mechanism of ozone cracking is studied by R. Criegee and released in 1957:

(1) Addition: ozone react with the double bond to form 1,2,3-trioxane.

$$>\!C\!=\!C\!<\; +\; \overset{O}{\underset{O\;\;\;O}{\triangle}} \longrightarrow\; >\!\underset{\underset{O}{|}}{C}\!-\!\underset{\underset{O}{|}}{C}\!<$$

(2) Decomposition: 1,2,3-trioxane decompose to a zwitterion and a carbonyl.

$$>\!\underset{\underset{O}{|}}{C}\!-\!\underset{\underset{O}{|}}{C}\!<\; \longrightarrow\; >\!C\!=\!\overset{O^-}{\underset{+}{O}}\; +\; >\!C\!=\!O$$

(3) Zwitterion react with ketone or dimerization or polymerization. If the reactivity of the carbonyl group is not very active, the zwitterion can form a dimer or be polymerized.

$$>\!C\!=\!\overset{O^-}{\underset{+}{O}}\; +\; >\!C\!=\!O\; \longrightarrow\; >\!\underset{\underset{O}{\diagdown\;\diagup}}{C\quad C}\!<$$

$$2>\!C\!=\!\overset{O^-}{\underset{+}{O}}\; \longrightarrow\; >\!\underset{\underset{O-O}{}}{C}\overset{O-O}{}C\!<$$

$$n>\!C\!=\!\overset{O^-}{\underset{+}{O}}\; \longrightarrow\; \{C\!-\!O\!-\!O\}_n$$

Chapter 4 Ozone degradation and stabilization

4.2 Stabilization of ozone degradation

According to the mechanism of ozone degradation, there are basically two kinds of methods for the stabilization of ozone degradation.

1. Physical protection

Wax is usually used as coating to protect rubber and corresponding product from ozone cracking. In practical application, the amount of wax is very important. If it is not enough, the coating cannot form, the protection is not efficient; if the wax is excess, it can detach from the surface very easily, which will lose the protection.

2. Chemical protection (antiozonant)

Addition of antiozonant is a chemical way to protect rubber from ozone cracking. There are several antiozonants: m-phenylenediamine, it is miscible with wax and can be used together with wax to achieve better protection. However, it is toxic, volatile, pollution; p-alkoxy-n-alkylaniline and N-substituent thiourea is also a good kind of antiozonants.

For the mechanism of antiozonant, L. D. Loan introduced the diversionary theory in 1968: the amine-containing antiozonant can react with the zwitterion as well as the ozonate heterocycle and therefore stabilize the process. It should be mentioned that antiozonant will not take effect immediately, it take time for the formation of zwitterion or other oxidized species.

$$\sim\!\!\!\sim\!\!\overset{\mathrm{H}}{\underset{\mathrm{H}}{\mathrm{C}}}\!\!=\!\!\overset{\mathrm{H}}{\underset{\mathrm{H}}{\mathrm{C}}}\!\!\sim\!\!\!\sim + O_3 \longrightarrow \sim\!\!\!\sim\overset{\mathrm{H}}{\underset{\mathrm{H}}{\mathrm{C}}}\!\!=\!\!O + \sim\!\!\!\sim\overset{\mathrm{O}^-}{\underset{\mathrm{H}}{\mathrm{C}}}\!\!-\!\!O_+ \xrightarrow{R_2NH} \overset{\mathrm{HOO}}{\underset{R_2N}{\diagdown}}\!\!\!\overset{}{\diagup}\!\!\overset{\mathrm{C}\sim\!\!\!\sim}{\underset{\mathrm{H}}{}}$$

$$\sim\!\!\!\sim\overset{\mathrm{O-O}}{\underset{\mathrm{H}\quad\mathrm{O}\quad\mathrm{H}}{\mathrm{C}\diagdown\!\!\!\diagup\mathrm{C}}}\!\!\sim\!\!\!\sim + R_2NH \longrightarrow \sim\!\!\!\sim\overset{\mathrm{H}}{\underset{}{\mathrm{C}}}\!\!=\!\!O + \sim\!\!\!\sim\overset{\mathrm{H}}{\underset{}{\mathrm{C}}}\!\!=\!\!O + R_2NOH$$

Further reading I: ozonation reactions and ozone cracking in elastomers

Both ozone and sunlight rapidly attack unprotected polymers which can significantly reduce the service life of a plastic. Particularly polymers with high unsaturation (e. g. rubbers) will suffer from ozone degradation, because the double bonds in unsaturated polymers readily react with ozone. However, ozone also reacts with saturated polymers but at a comparatively slower rate. The reaction of ozone with double bonds causes chain scission. The general mechanism of ozone degradation is shown below:

$$\begin{array}{c}\diagdown\!=\!\diagup + O_3 \longrightarrow \overset{O}{\underset{O\ O}{\diagdown\!\!\!\diagup}} \longrightarrow \overset{O^+-O^-}{\underset{/\ \ \backslash}{CH}} + \diagup\!\!=\!O \\[1em] \longrightarrow \overset{O-O}{\underset{O}{\diagdown\!\!\!\diagup}} + -\overset{|}{\underset{|}{C}}-O-O + \overset{O-OH}{\underset{OR}{\diagdown\!\!\!C\!\!\diagup}} \end{array}$$

Chain scission and oxidation causes a decrease in cross-link density (elastomer), or molecular weight and molecular weight distribution (thermoplast), and a change in composition (oxidation). The result is a more or less steady decline in the (mechanical) properties.

The aging is greatly accelerated by stress; one usually observes surface cracks in the direction perpendicular to the applied strain when a critical stress value is exceeded. At rather low stress values just above the critical value, long and deep cracks are observed, whereas at high stress values, the ozone cracks become more numerous and are finer in size. The microscopic disintegration of the surface causes dulling and a bluish sheen of the surface of rubber goods. This phenomenon is known as "frosting" because it often resembles actual frost. It is greatly accelerated by humidity and heat and is most noticeable on the bright finish of air-cured rubber goods. Frosting can be avoided or reduced by certain types of high melting point waxes or by antiozonants such as *para*-phenylenediamines (PPDs) and derivatives thereof (Dialkyl PPD).

In general, the resistance to ozone cracking and frosting will depend on the chemical composition of the polymer. Elastomers are particularly susceptible to ozone attack, particularly those with electron donating side groups (e. g. methyl groups in isoprene), whereas rubbers with electron-withdrawing side groups (e. g. chlorine in neoprene) are noticeably less susceptible to ozone attack due to the deactivating effect of the halogen on the double bond.

The degree of ozone degradation will depend on the composition of the atmosphere and temperature. Usually, the ozone concentration is rather low. Nevertheless, even low values of ozone at ambient temperatures can cause significant degradation over time.

Further reading Ⅱ : antiozonants
para-phenylenediamines

Ozone and sunlight rapidly attack unprotected polymers which can significantly reduce their service life. Particularly polymers with high unsaturation, e. g. rubbers, are prone to suffer from ozone degradation because the double bonds in unsaturated polymers readily react with ozone. To prevent or to slow down ozone initiated oxidative degradation, antiozonants are frequently added. The most powerful and most common class of antiozonants are

Chapter 4 Ozone degradation and stabilization

para-phenylene diamines (PPDs) which have the general structure:

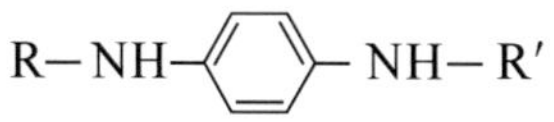

R—NH—〈 〉—NH—R'   Type I : N,N'-dialkyl-*p*-phenylenediamine

p-phenylenediamine(PPD)   Type II : N-alkyl-N'-aryl-*p*-phenylenediamine

Type III : N,N-diaryl-*p*-phenylenediamine

PPDs are not only efficient antiozonants but are also very effective primary antioxidants. Their reactivity and efficiency depends on the substituents on the nitrogen. In general, dialkyl substituted amines are the most reactive PPDs followed by arylalkyl-substituted and bisaryl-substituted amines; probably because the N—H bond of alkyl substituted PPDs has a lower bond dissociation enthalpy than those of aryl substituted amines. However, the activity of the different types of PPDs also depends on their solubility in the rubber, the temperature, and aging conditions. For this reason, blends of PPDs are often used. In some cases, the efficiency of antiozonants can be further increased by incorporation of waxes and certain synergistic antioxidants. All three types of PPDs are effective antiozonant in natural rubber and polyisoprene because they are only moderately soluble in these rubbers, and thus can migrate to the rubber surface to provide good ozone protection. Diaryl PPDs, on the other hand, they are usually more effective in polychloroprene than the others.

There are at least three competing mechanism of ozone protection:

(1) The antiozonants react faster with ozone than the rubber and, therefore, acts as ozone scavengers.

(2) The ozone-antiozonant reaction products form a protective film on the rubber surface preventing ozone from reacting with the rubber.

(3) The antiozonants react with the radical sites of the rubber fragments, forming new cross-links and, thus, restore the rubber network.

The reaction between N-alky-substituted *para*-phenylenediamines and ozone is shown below:

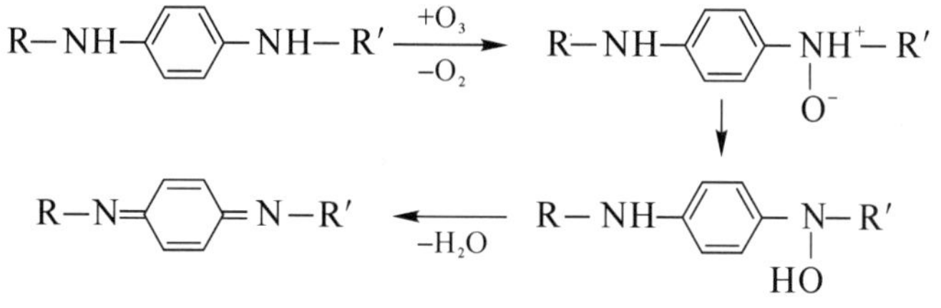

$$R-NH-\langle\ \rangle-NH-R' \xrightarrow[-O_2]{+O_3} R-NH-\langle\ \rangle-NH^+-R'$$
$$\downarrow O^-$$
$$R-N=\langle\ \rangle=N-R' \xleftarrow{-H_2O} R-NH-\langle\ \rangle-N-R'$$
$$HO$$

Ozone oxidizes the amino group of the phenylenediamine yielding a hydroxylamine. Subsequent elimination of water yields quinonediimine which can be further oxidizes by peroxyl radicals to produce nitrones and dinitrones (not shown).

Waxes provide effective protection under static load. However, protection under dynamic loads is possible only with antiozonants.

Chapter 5 Microbiological degradation and stabilization

In principle, organic molecules from natural resources including biopolymers such as cellulose and pectin can be decomposed by microorganisms through metabolization mechanism. However, the microorganisms show stress-tolerance for synthetic molecules (xenobotics) due to the lack of possible metabolization mechanism.

5.1 Basics for biodegradation

As known to all, biodegradation of natural polymers derived from the specific enzyme protein. These enzymes locate at the cell wall or in the plasma. Enzyme shows highly catalytical activity and high specification, and should be used in mild condition or physiological condition. Generally, there are two types of biodegradation for polymers: ① release enzyme to the outside medium and decompose the polymeric substrate to assimilate molecules. ② close contact with polymers to facile the enzyme catalysis reactions.

As mentioned in previous chapters, biodegradation is actually the enzyme catalyzed hydrolyzation. Table 5.1 shows the typical hydrolyzation for some common polymers.

Table 5.1 Typical hydrolyzation for some common polymers

Polymers	Hydrolyzation
C—C polymers	$(CH_2-C(CN)_2)_n + H_2O \to nCH_2O + nCH_2(CN)_2$ $(CH_2-C(CN)COOR)_n + H_2O \to nCH_2O + nCN \to CH_2COOR$
Polyimide, PMMA	$(CH_2-NH-CO)_n + H_2O \to nNHCH_2COOH$
Polyester, PC	$(CH_2-O-CO)_n + H_2O \to HOCH_2COOH$
POM	$(CH-O-CH_2)_n + H_2O \to nROM + nHOCH_3CH_2OHOR$
Inorganic polymers	$(P-N)_n + H_2O \to 2nROH + nNH_2 + nP(OR)_2$

Due to the specific recognition between enzyme and substrate, the biodegradation mechanism is different for different kinds of polymers.

1. Starch

Starch can be decomposed by microorganism through the hydrolyzation or digestion. Chemically, starch is synthesized during the photosynthesis and stored as particles. The

Chapter 5 Microbiological degradation and stabilization

starch is composed by D-glucose, while linear starch is connected by α-1, α-4 glycosidic bond and branched starch is connected by α-1, α-6 glycosidic bond. Linear starch locates inside of the particle and can be swollen in water; branched starch is inside of the particle and can be soluble in hot water, but insoluble in cold water. The degree of polymerization for linear starch is ca. 200 - 1,000, and branched one is 6,000 - 280,000.

Most of the microorganisms can decompose starch, no matter if they are aerobic oranaerobic. Amylase is the specific enzyme for starch. Aspergillus niger or Aflatoxin on the surface of starch can generate amylase. Normally, branched starch is more easily to be decomposed than the linear one, which is due to the presence of phosphorous in the branched starch.

2. Cellulose

Cellulose is the major component for cell membrane of plants. The cellulose molecule is a linear polymer composed by D-glucose connected by 1,4-glycosidic bond. The mean DP for cellulose is ca. 5,000 - 6,000, with molecular length of 2.5 - 3.0 μm. Cellulose is composed by microcrystalline and noncrystalline parts.

Most of the microorganisms can decompose cellulose, no matter if they are aerobic oranaerobic. Actually, the decomposition mechanism of cellulose is different for different celluloses. Manbdels and Reese isolated different enzymes from cellulose from trichoderma viride: C_1 and C_x cellulose. Selby isolated two different enzymes from C_x cellulose through gel filtration technology: carboxymethyl cellulose and cellobiase. It is found that single C_x enzyme cannot decompose cotton cellulose, but can decompose carboxy methyl cellulose.

3. Natural rubber

Natural rubber can be decomposed by microorganisms. Nickerson isolated black yeast-like fungus and pink bacterial through percolation of the vulcanization product of natural rubber. These microorganisms can induce pores on the surface of natural rubber. The surface of natural rubber can be decomposed into small particles, while there are exoenzymes on the surface of these particles to induce further degradation.

Natural rubber can be oxidized by oxidants. It is believed that the microorganism will attack and oxidize the chain end to achieve degradation.

4. Polyolefins

Normally, polyolefins cannot be decomposed by microorganisms due to the large molecular weight and hydrophobic nature of polyolefins. Upon the introduction of hydrophilic group or oxidize some of the polymeric structures, enzyme can easily penetrate polymers to achieve biodecomposition. For example, partially oxidized or hydrolyzed PS can be decomposed by penicillium and moraxella lacunata due to their hydrophilic nature.

5. Polyurethane, polyimide and polyester

Synthetic polymers with urethane, imide and ester bonds are very similar to natural polymers, and can be decomposed by microorganisms.

The structure of polyurethane is similar to polypeptide and can be decomposed by microorganisms.

Polyimide prepared from aminoacid shows good biocompatibility and biodegradation property, which shows potential application in biomedical areas. Polyhydroxy propylmethylacrylamide can be hydrolyzed by papain.

Polyketide ester with low molecular weight can be decomposed by aspergillus niger and aflatoxin. Introduction of methyl group into the polymeric structure can improve the stability toward biodegradation.

6. Plasticizer

Polymer, such as PVC, is inert toward microorganism. However, the plasticizer in the PVC is sensitive to biodegradation, which will induce the decomposition of PVC products. Meanwhile, plasticizer can also be decomposed by enzyme at the interface of soil and plastic, which will lose the toughness of the plastics.

5.2 Stabilization of biodegradation

As shown above, the biodegradation process involves the enzyme catalyzed reaction. The activity of enzyme is affected by the humidity, temperature and pH of the environment. Therefore, we can improve the biodegradation stability of polymers through the control of the metabolization process of microorganism by both chemical and physical methods.

1. Chemical modification

To limit the attachment of microorganism with polymer, the increasing of hydrophobic property of polymer can be used to improve their stability toward biodegradation. For example, modification of cellulose with different groups and tuning their degree of substitution can improve the stability of cellulose.

2. Stabilizer for biodegradation

Due to the limitation of chemical modification, stabilizer is widely used to improve their stability toward biodegradation. The stabilizer should have the following characteristics:

(1) Broad spectrum antibacterial property.
(2) High anti-bacterial activity.
(3) Long term anti-bacterial property.
(4) No side effect, such as coloring, photo and thermal stability.
(5) Good miscibility with the additives such as plasticizer, antioxidant, thermal stabilizer and different kinds of auxiliaries.
(6) Non-corrosive, no smell.
(7) Good thermal stability.
(8) Good stability during storage.
(9) Easy to use.
(10) Harmless to human and environment.

Chapter 6   Characterization of polymer degradation

A few effects of outdoor weathering, such as discoloration and embrittlement, are readily detectable by visual or manual inspection. Quantitative investigations nevertheless require modern analytical techniques capable of probing material changes at different levels, from determining macroscopic bulk properties to understanding molecular mechanisms. The hierarchy of methodologies which has been applied to degradation investigation is summarized in Table 6.1.

Table 6.1   Hierarchy of degradation investigative methodologies

Scale	Aspect investigated	Methodology
Macroscopic	Decrease in mechanical properties with time	Tensile test, fracture energy measurements
Microscopic	Development of surface microcracks	Optical and electronic microscopy
Supramolecular	Change in morphology	Differential scanning calorimetry(DSC), transmission electron microscope(TEM)
Macromolecular	Change in molecular weight distribution	GPC
Chemical	Stable degradation products	Oxygen uptake, Fourier transform infrared spectrometer(FTIR), Raman, Nuclear magnetic resonance(NMR), impedance spectroscopy
Reactive intermediates	Free-radical formation	ESR, chemiluminescence

Many analytical techniques used to inspect the cited properties are common to the field of polymer characterization: vibrational spectroscopy (FTIR, Raman), magnetic resonance spectroscopy (NMR, ESR) and liquid chromatography [GPC, High-performance liquid chromatography (HPLC)]. A few methods, such as oxygen consumption and chemiluminescence, are more specific to oxidative degradation. Mechanical tests are frequently used in combination with other analytical tools to asset the effects of degradation on mechanical properties.

6.1 Mechanical tests

Many plastic materials are used for their load-bearing properties, and mechanical testing occupies a dominant position among the different criteria for degradation. Typical molecular consequences of degradation processes are chain scission and crosslinking. These effects have a profound influence on the mechanical properties of the material. In general, a predominance of crosslinking over chain scission increases the tensile strength while decreasing the elongation at break and the viscoelastic flow of the material. The reverse effects are observed for chain scission. The loss of toughness in semicrystalline polymers arises from oxidative chain scission in the amorphous region, which involves:

(1) scission of tie molecules.

(2) recrystallization of lower MW polymer fraction following scission.

(3) densification from increasing polarity.

In thermal aging of polypropylene, for instance, a rapid transition from ductile to brittle behavior was observed before the appearance of carbonyl groups in the FTIR spectra. Recent data nevertheless indicate that some of the delay in C=O build up may result from the loss, during the initial degradation stage, of volatile ketones and carboxylic acids which hence elude detection by FTIR. Gel permeation chromatography reveals the presence during the induction period of chain scission, most likely b-scission of alkoxy radicals, which accompanies loss in mechanical strength when the weight-average molecular weight M_w drops below some value of the order of 105 g/mol (see Fig. 6.1).

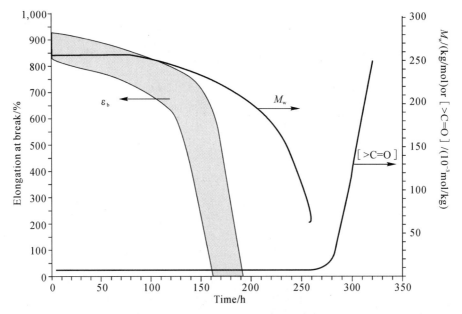

Fig.6.1 Change in elongation at break, weight-average MW (M_w), and carbonyl content as a function of degradation time, in the thermo-oxidative degradation of unstabilized *i*-PP

Chapter 6 Characterization of polymer degradation

Degradation, particularly photodegradation, is a surface phenomenon which affects essentially only the outer few hundred microns from the surface. Scanning electron microscopy invariably reveals the appearance of multiple cracks and microcracks at the surface of degraded polymers, as shown in Fig. 6.2.

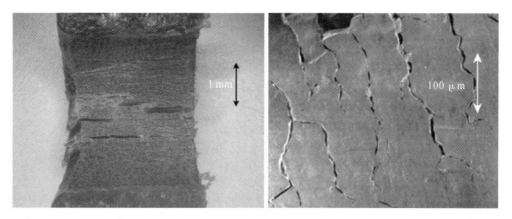

Fig.6.2 Thermally degraded polypropylene samples: left for macroscopic aspect, showing the interior of a water pipe degraded in hot water (six months, 90 ℃); right for microscopic (SEM) view of an HAS-stabilized sample thermally aged in an oven (54 days at 135 ℃)

The morphology of these cracks depends on the type of polymer and the degradation conditions. Crack formation normally occurs on unstrained material according to the following sequence: oxidation of the superficial layer → change in material density → volume contraction → differential deformation → crack formation.

Oxidized polymers show an increase in density owing to the presence of polar functions. Densification of the degraded material creates internal stresses between the outer layer and the intact interior. Ultimately, cracks will appear when the mechanical strength of the degraded layer falls below the differential tensile stresses. Water absorption from ambient humidity or from rain, followed by surface evaporation, is another source of cyclic mechanical stresses which can favor crack formation in certain polymers. The spatial irregularities in oxidation are another source of differential stresses between degraded and undegraded domains.

Fracture usually occurs from surface cracks. Starting from the degraded layer, these surface microcracks can easily propagate into the internal intact layer by stress concentration at the crack tip. Some mechanical tests are more sensitive to the outer layers than to the bulk properties, and this peculiarity should not be overlooked when interpreting experimental results. Tensile strength of degraded elastomers, for instance, shows a different behavior when compared with tensile elongation, because the former depends on the entire cross-section of the material, whereas ultimate elongation is correlated with surface modulus because cracks initiated at the hardened surface immediately propagate through the sample.

6.2 Gel permeation chromatography

Polymer degradation involves cleavage of bonds, resulting in a decrease in MW, branching, cyclization, and crosslinking. These changes in the MW and MWD, and their evolution with the extent of degradation, can give valuable insights into the degradation mechanism. Before the advent of gel permeation chromatography (GPC), different MW averages such as M_n, M_v or M_w, were used routinely in evaluation of polymer degradation processes. Determination of the number-average MW (M_n) has the major advantage that it is directly related to the inverse of the number of molecular chains per unit polymer mass [following equation, where n_i is the polymer molar fraction of molecular mass M_i].

$$M_n = (\sum n_i \cdot M_i)$$

In polymer degradation experiments, it is convenient to define the scission indexs as the number of broken bonds per initial macromolecule. Because each mainchain scission produces an additional molecule, the indice s is directly given by the change in number-average MW regardless of the initial MWD and degradation mechanism [following equation, with n representing the number of polymer chains per gram of polymer, and the superscript "o" referring to the initial conditions].

$$s = [n - n°]/n° = [M_n°/M_n]^{-1}$$

Pure chain scission seldom occurs without other competitive radical reactions, except perhaps during the early stage of oxidative degradation. When chain scission and crosslinking happen simultaneously, the polymer molecular weight may increase or decrease depending on the relative importance of each process. For random scission and crosslinking, Flory and Charlesby have shown that a three dimensional network begins to occur at the "gel point", when the crosslink density reaches one unit per weight average molecule. The soluble fraction (s) after the gel point can be used to determine the relative yield of chain scission to crosslinking during degradation.

GPC, particularly in its multidetection version, is now a mature and well-accepted technique for MW characterization. The capability of GPC to determine changes throughout the MWD, in addition to the different MW averages, opens new possibilities for polymer degradation studies. The scission probability as a function of position along the molecular chain, for instance, could be inferred from the MWD of degraded polymer.

In a given polymer where most of the covalent bonds have on the average the same bond energy, it is expected that the rate of bond scission may not depend on its position along the chain (random scission). Although this is generally true for the thermal and photodegradation of several polymers, other bond scission statistics can also be observed, as shown in Fig. 6.3.

The Gaussian scission probability distribution function, with a preference for mid-chain scission, is frequently encountered in shear-induced mechanochemical degradation. The

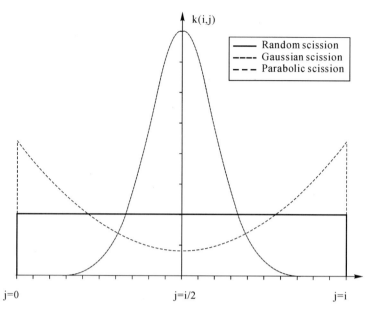

Figure. 6.3 Some commonly encountered scission probability distribution functions

parabolic distribution, on the contrary, indicates a preference for chain-end degradation. This phenomenon has been reported in the hydrolysis of dextran, as a result of chain branching.

A fourth situation, not depicted in Fig. 6.3, is encountered in some polymers which have weak links at a chain end or a low ceiling temperature. In the former situation, initiation starts at the chain end bearing the weak bond with volatilization of a monomer out of the reaction medium. In the second case, exemplified by polyacetals, initiation also occurs at a terminal position, but the process continues and the monomers keep being evolved until complete volatilization of the polymer (unzipping).

6.3 Fourier transform infrared spectroscopy

Identification of the degradation mechanism, and its evolution with time, are mainly based on analysis of the intermediate and final chemical products. FTIR proves to be particularly suited to that purpose. The infrared absorption intensity A_s is related to the square of the change in dipole moment (μ) during molecular vibration according to following equation, where ν is the frequency of the band center, and Q, the normal coordinate of the vibration.

$$A_s \propto \nu \cdot (\partial \mu / \partial Q)^2$$

The dynamic dipole moments of oxidized compounds are high, owing to the polar nature of the oxidation products, and FTIR provides a sensitive and quantitative technique for the investigation of oxidative degradation. Oxidized polymers show changes in the IR absorption spectrum over the whole range of measuring wavelengths. Most noticeable are the appearance of complex overlapping bands in the carbonyl (1,600-1,800 cm^{-1}) and hydroxyl (3,200 – 3,600 cm^{-1}) regions (see Fig. 6.4).

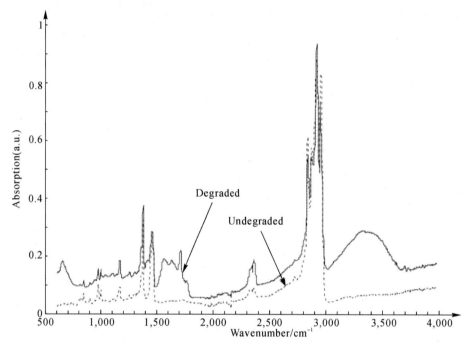

Fig. 6.4 Evolution of the ATR-FTIR spectra of an unstabilized commercial i-PP, before (—) and after (- - --) 3 h of thermal treatment at 120 ℃

The content of carbonyl groups is an indirect indicator of the number of scission events. The average ratio of the absorption coefficient of hydroxyl (—O—H) to carbonyl (C=O) functions is approximately 3.4 : 1, and this value can be used to estimate the relative proportion of hydroperoxides to other oxidation products of degradation.

IR absorption bands in amorphous solids are rather broad, with extensive overlap. Direct identification and quantification of the numerous oxidation products, most often of similar chemical structures, in a degraded polymer are difficult. More precise conclusions about the nature of absorbing species can be obtained by selective chemical derivatization. Selective modification of functional groups with reactive gases, such as SF_4, NH_3, SO_2, or NO, results in a shift in absorption band positions, which can then be compared with model compounds to allow for a better chemical assignment of the absorbing species (see Table 6.2).

Chemical derivatization can also be used for structure elucidation in complex mixtures. For instance, dimethyl sulfide has been used to differentiate peracids from other peroxy compounds. The rate of conversion of dimethyl sulfide into dimethyl sulfoxide is almost instantaneous with peracids equation (6 - 1), whereas it is slow with *sec*- and *tert*-

hydroperoxides equation(6-2).

$$R_2C(O-OH)(O) + CH_3-S-CH_3 \rightarrow R_2C(OH)(O) + CH_3-S(=O)-CH_3 \qquad (6-1)$$

$$R_2CH-O-OH + CH_3-S-CH_3 \rightarrow R_2CH-OH + CH_3-S(=O)-CH_3 \qquad (6-2)$$

Table 6.2 Some common derivatization techniques for identification by FTIR

Functional group	Derivatization reaction	IR frequency/cm^{-1}
Carboxylic acid	$R-C(=O)-OH + NH_3 \rightarrow R-COO^-$	1,564
Carboxylic acid	$R-C(=O)-OH + SF_4 \rightarrow R-C(=O)-F$	1,848 1,841(α-branching)
Hydroperoxide	$R-O-OH + NO \rightarrow R-ONO_2$	1,276 + 1,290
	The sharp and intense absorption bands of nitrates and nitrites can be used to differentiate primary from secondary and tertiary compounds	
Alcohols	$R-OH + NO \rightarrow R-NO_2$	778
Aldehydes	$R-CH=O + 2Ag(NH_3)_2OH \rightarrow RCOO^- NH_4^+$ $+ 2Ag\downarrow + 3NH_3\uparrow + H_2O$ (Tollens' reagent does not react with ketones)	1,550 – 1,555

The carboxyl group, in the presence of hydroxyls, may be differentiated by reacting with silver trifluoracetate equation(6-3) or with trifluoroanhydride equation(6-4),(6-5).

$$-COOH + CF_3COOAg \rightarrow -COOAg + CF_3COOH \qquad (6-3)$$
$$-OH + (CF_3CO)_2O \rightarrow -O-CO-CF_3 \qquad (6-4)$$
$$-COOH + (CF_3CO)_2O \rightarrow -COO-CO-CF_3 \qquad (6-5)$$

6.4 Magnetic resonance spectroscopy

1. Nuclear magnetic resonance (NMR)

NMR is an absolute and unique method of resolving the microstructure of polymer species. This technique has been instrumental in the elucidation of degradation mechanisms in several polymer systems, including polyvinyl chloride (PVC).

At around its glass transition temperature (81 ℃), unstabilized commercial PVC starts to discolor with release of hydrochloric acid, according to the global reaction:

$$+H_2C-CHCl+_n \longrightarrow +HC=CH+_n + n\,HCl$$

Because the thermal stability of PVC is substantially lower than one may expect from its nominal structure, numerous studies have been initiated to identify the defect sites responsible for this anomaly. After considerable controversy concerning the role, nature, and importance of possible irregular structures, the most active sites for initiating PVC degradation were finally identified by ^{1}H- and ^{13}C NMR as internal allylic [see Fig. 6.5 (Ⅰ),(Ⅱ)]and tertiary chlorine structures [see Fig. 6.5 (Ⅲ),(Ⅳ)].

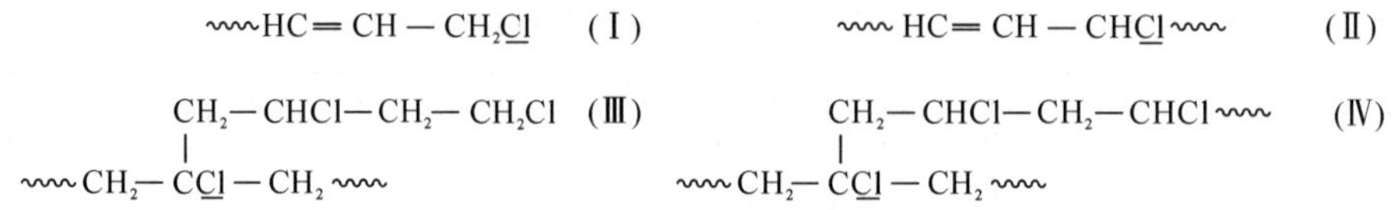

$$\sim\sim HC=CH-CH_2\underline{Cl} \quad (Ⅰ) \qquad \sim\sim HC=CH-CH\underline{Cl}\sim\sim \quad (Ⅱ)$$

$$\begin{array}{c} CH_2-CHCl-CH_2-CH_2Cl \quad (Ⅲ) \qquad CH_2-CHCl-CH_2-CHCl\sim\sim \quad (Ⅳ) \\ | \qquad\qquad\qquad\qquad\qquad\qquad | \\ \sim\sim CH_2-C\underline{Cl}-CH_2\sim\sim \qquad\qquad \sim\sim CH_2-C\underline{Cl}-CH_2\sim\sim \end{array}$$

Fig.6.5　Major labile structural defects in emulsion PVC

It was believed for a long time that head-to-head radical addition to monomers is a major route for formation of labile structures. Kinetic studies, in association with NMR measurements, reveal that formation of internal allylic and tertiary chlorine structures actually proceeds through an intramolecular or intermolecular chain transfer reaction to polymer [equation(6-6),(6-7); VC = vinyl chloride].

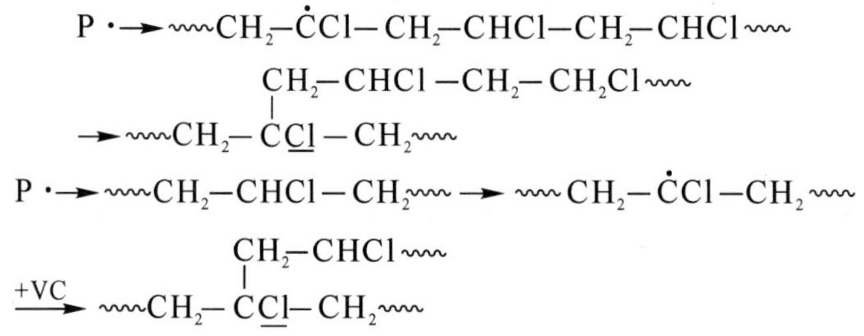

$$\begin{array}{c} P\cdot \longrightarrow \sim\sim CH_2-\dot{C}Cl-CH_2-CHCl-CH_2-CHCl\sim\sim \\ CH_2-CHCl-CH_2-CH_2Cl\sim\sim \\ | \\ \longrightarrow \sim\sim CH_2-C\underline{Cl}-CH_2\sim\sim \end{array} \quad (6-6)$$

$$\begin{array}{c} P\cdot \longrightarrow \sim\sim CH_2-CHCl-CH_2\sim\sim \longrightarrow \sim\sim CH_2-\dot{C}Cl-CH_2\sim\sim \\ CH_2-CHCl\sim\sim \\ \xrightarrow{+VC} \sim\sim CH_2-C\underline{Cl}-CH_2\sim\sim \end{array} \quad (6-7)$$

2. Electron spin resonance (ESR)

Most mechanisms for the degradation of polymers during processing, storage, and long-term use include free radicals. The technique of ESR is therefore a logical choice for the study of polymer degradation. Two kinds of free radicals are of key importance in determining degradation kinetics: ① the carbon-centered macroalkyl C—C·—C. ② the oxygen-centered alkylperoxyl ROO·, alkoxyl RO·, and acylperoxyl R(CO)OO· radicals. Most of these species have unique ESR spectra which allow their unambiguous identification under favorable conditions. In addition to structure identification, the decay signal of the radicals with time provides information about termination rates and reaction orders.

At room temperature and in solutions, free radicals are too reactive, and their concentrations too low, to be conveniently detected by ESR spectroscopy. To observe and quantify transient radicals, it is customary to transform the unstable species into stable nitroxide radicals with the spin-trapping technique. Spin-trapping of a macroalkyl radical with pentamethy lnitrosobenzene, for instance, can be represented schematically by following equation. The structure of the parent radical could be deduced from the ESR spectrum of the spin adducts, using standard rules of spin coupling and experimental coupling constants.

$$\underset{\underset{H_3C}{H_3C}}{\overset{\overset{H_3C}{H_3C}}{\bigcirc}}-N=O + \cdot R \longrightarrow \underset{\underset{H_3C}{H_3C}}{\overset{\overset{H_3C}{H_3C}}{\bigcirc}}-\underset{\underset{O\cdot}{|}}{N}-R$$

6.5 Oxygen uptake

Oxygen uptake provides a simple, yet highly sensitive means to monitor oxidative degradation. A weighed amount of polymer is hermetically enclosed with a fixed amount of oxygen under controlled pressure. Oxygen consumption can be monitored by measuring volumetric change, or the decrease in oxygen content by gas chromatography. The first method cannot discriminate oxygen consumption from the release of volatile gases (CO, CO_2, H_2O, etc.) and tends to underestimate the amount of oxygen consumed. The second technique benefits from the high sensitivity and separation capability of gas chromatography which, with proper care, allows determination of oxygen consumption rates down to $10-13/[mL/(g \cdot s)]$. The gas chromatography approach has been used to determine the instantaneous oxidation rate of elastomers even at room temperature, and reliable lifetime can be predicted under ambient conditions without having recourse to extrapolation from accelerated testing.

Measurement of oxygen consumption is the starting point for several analyses of polymer oxidation kinetics. The oxygen uptake curve for most polyolefins shows an induction period with a very slow oxidation rate, before reaching a "steady state" with constant slope. The value of this slope has been often interpreted as the limiting oxidation rate of the polymer. The induction period is generally much shorter in photo degradation than in thermal degradation (see Fig. 6.6). The main difference stems from the different modes of initiation. The activation energy of primary photochemical processes, 8 - 40 kJ/mol, is generally much less than the energy required for thermal activation (120 - 200 kJ/mol). Because photo degradation generally occurs at low temperature, photo decomposition of peroxides is slow and this class of products tends to accumulate rather than being decomposed to initiate chain branching reactions as in thermal degradation.

During oxygen uptake, part of the absorbed oxygen is transformed into hydroperoxides, while the remaining fractions are found in hydroxyl- and carbonyl-containing compounds formed by competitive reactions of P· and POO· radicals. The relative importance of each process depends on the oxygen partial pressure, and various kinetic equations have been derived to relate the rate of oxygen absorption ($d[O_2]/dt$) to the formation of oxidation products, under low and high oxygen pressure conditions. Kinetic information, and rate coefficients for propagation and termination, have frequently been obtained from the fit of the oxygen consumption curve at relatively low conversions. This procedure, however, is complicated by the facts that intermediate oxidation products may be orders of magnitude more susceptible to oxidation than the starting material, and that degradation in solid polymers is fundamentally heterogeneous. Therefore, any obtained value can yield only average information about the degradation process.

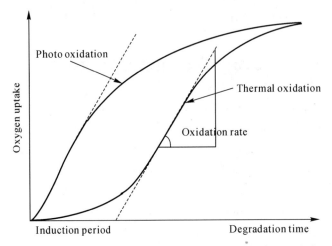

Fig.6.6 Oxygen uptake curves during photo-oxidation and thermal oxidation of unstabilized polymers

6.6 Chemiluminescence

As with oxygen micro-uptake, chemiluminescence is an ultrasensitive technique for determining the rate of oxidation at the earliest stage of degradation. Oxidative degradation of polymers is generally accompanied by weak photon emission, with a very low quantum yield (approximately 109). The light-emitting reaction which accompanies free-radical oxidation is generally attributed to an exoenergetic termination reaction of peroxy radicals according to the Russell mechanism equation(6 - 8).

$$\begin{array}{c}\underset{R'}{\overset{R}{>}}\underset{H}{\overset{O}{\underset{|}{C}}}\underset{O\cdot}{\overset{O\cdot}{\underset{|}{O}}}\underset{R}{\overset{}{}} \longrightarrow \underset{R'}{\overset{R}{>}}\underset{H}{\overset{O^*}{\underset{|}{C}}}\underset{O}{\overset{O}{\underset{|}{\|}}}\underset{R}{\overset{}{}} \\ \longrightarrow \underset{R'}{\overset{R}{>}}C=O^* + {}^1O_2 + ROH \\ H = -460 \text{ kg/mol} \end{array}$$

(6 - 8)

The Russell mechanism requires one of the terminating radicals to be either primary or secondary so that a six-membered transition state can be formed. Such a mechanism may be prevalent in PE and PA, but not in PP, where the chain carrying radical is tertiary. For this polymer, more complex alternative routes for chemiluminescence-producing reactions have been proposed.

Further reading Ⅰ: thermogravimetric analysis of polymers

Thermogravimetric analysis (TGA) is one of the most popular analysis techniques to study the decomposition process of polymeric materials in controlled atmospheres at various temperatures.

Thermal degradation usually produces volatile compounds. Thus, measuring the weight loss at different temperatures provides information about the thermal stability of polymeric materials. The onset of weight loss is often used to define the upper temperature limit of thermal stability. However, in many cases degradation has already taken place without a detectable weight loss, for example by chain scission or cross-linking reactions. Thus, in some cases the upper service temperature is noticeably lower than the onset of weight loss. Nevertheless, thermogravimetric methods are often an excellent choice to study thermal degradation processes of polymers. Both isothermal and dynamic thermogravimetric methods are employed. In the case of a dynamic TGA, the weight loss of a sample is recorded while the temperature is continuously increased at a constant rate in a controlled atmosphere whereas an isothermogravimetric analysis (IGA) meausures the weight loss at a constant temperature as a function of time and atmosphere.

For a reaction of the type:

$$A(solid) \rightarrow B(solid) + C(gas)$$

The rate of weight loss, e.g. the amount of gas released will depend on the heating rate and temperature and can be described with an empirical equation of the form:

$$R_d = d\alpha/dt = k(T) \cdot f(\alpha)$$

Where k is the rate constant of the decomposition process, $f(\alpha)$ is an (empirical) reaction model and α is the extend of conversion which is defined as:

$$\alpha = \frac{M_0 - M(t)}{M_0 - M(t \rightarrow \infty)}$$

Where M_0 is the initial weight and $M(t)$ is the weight at time t.

A popular equation to describe the decomposition kinetic is:

$$f(\alpha) = c \cdot \alpha^m \cdot (1-\alpha)^n \cdot [\ln(1-\alpha)]^p$$

with m, n, and p parameters that depend on the reaction mechanism. They are usually determined by model fitting.

Table 6.3 shows the reaction models for thermal decomposition in solids.

Table 6.3 Reaction models for thermal decomposition in solids

Mechanism	$f(\alpha)$	$g(\alpha)$
Power Law 1	$4\alpha^{3/4}$	$\alpha^{1/4}$
Power Law 2	$3\alpha^{2/3}$	$\alpha^{1/3}$

Continued table

Mechanism	$f(\alpha)$	$g(\alpha)$
Power Law 3	$2\alpha^{1/2}$	$\alpha^{1/2}$
First Order	$(1-\alpha)$	$-\ln(1-\alpha)$
Second Order	$(1-\alpha)^2$	$(1-\alpha)^{-1}-1$
Third Order	$(1-\alpha)^3$	$[(1-\alpha)^{-2}-1]/2$
Diffusion 1-D	$1/2\alpha^{-1}$	α^2
Diffusion 2-D	$[-\ln(1-\alpha)]^{-1}$	$(1-\alpha)\ln(1-\alpha)+\alpha$
Diffusion 3-D		$[1-(1-\alpha)^{1/3}]^2$
Contracting area	$2(1-\alpha)^{1/2}$	$1-(1-\alpha)^{1/2}$
Contracting volume	$3(1-\alpha)^{2/3}$	$1-(1-\alpha)^{1/3}$
Avarami-erofeev (A2)	$2(1-\alpha)[-\ln(1-\alpha)]^{1/2}$	$[-\ln(1-\alpha)]^{1/2}$
Avarami-erofeev (A3)	$3(1-\alpha)[-\ln(1-\alpha)]^{2/3}$	$[-\ln(1-\alpha)]^{1/3}$
Avarami-erofeev (A4)	$4(1-\alpha)[-\ln(1-\alpha)]^{3/4}$	$[-\ln(1-\alpha)]^{1/4}$

The rate constant k describes the relationship between the reaction rate $(d\alpha/dt)$ and the extend of conversion α. Its temperature dependence can be described with an Arrhenius equation:

$$k(T) = A \cdot \exp(-E_a/RT)$$

where R is the gas constant and T is the temperature. The two parameters E_a and A are the activation energy and the frequency factor of the reaction. Combining the equation of $k(T)$ and R_d gives:

$$d\alpha/dt = k(T) \cdot f(\alpha) = A \cdot \exp(-E_a/RT) \cdot f(\alpha)$$

Integration of the expression $d\alpha/dt = k(T) \cdot f(\alpha)$,

$$g(\alpha) = \int_0^\alpha f(\alpha)^{-1} d\alpha = k(T)t$$

and substitution for $k(T)$ yields after rearranging,

$$g(\alpha)/t = A \cdot \exp(-E_a/RT)$$

This expression is often written in its logarithmic form:

$$\ln t = \ln[g(\alpha)/A] - (E_a/RT)$$

Choosing a fixed value of α, the first term on the right hand side of this equation is a constant. Then apparent activation energy E_a can be determined from the slope of a log t against $1/T$ plot.

The equation above can also be expressed as a function of heating rate $\beta = dT/dt$:

$$\beta \cdot d\alpha/dT = A \cdot \exp(-E_a/RT) \cdot f(\alpha)$$
$$\ln(\beta \, d\alpha/dT) = \ln[A/f(\alpha)] - E_a/RT$$

Thus measuring the temperature T and the conversion rate $d\alpha/dT$ for a fixed extend of conversion α at different heating rates β and plotting $\ln(\beta d\alpha/dT)$ versus $1/T$ yields the activation energy E_a. This is oconversional method has been suggested by Friedman.

Chapter 6 Characterization of polymer degradation

Many other approximate equations for the calculation of the apparent activation energy have been suggested. Two of the most popular methods are:

Flynn-Wall-Ozawa:

$$\ln\beta = \ln[AE_a/g(\alpha)R] - 5.331 - 1.052\, E_a/RT$$

Coats-Redfern

$$\ln[g(\alpha)/T^2] = \ln[AR/\beta E_a] - E_a/RT$$

Both methods are based on using approximations for the expression:

$$g(\alpha)\int_0^\alpha f(\alpha)^{-1}d\alpha = \frac{A}{\beta}\int_{T_0}^T \exp(-E/RT)dT = k(T)t$$

The accuracy of these methods depends on the extend of conversion α. Usually conversions between 2% to 20% give results of reasonable accuracy.

The effective activation energy of thermal-oxidative degradation is often a good indicator for the heat stability of a plastics. However, the activation energy is not always a constant but often depends on the extend of reaction as well as on the atmosphere. In the case of PE and PP, values between about 150 kJ/mol and 250 kJ/mol have been measured under nitrogen whereas polystyrene (PS) has a practically constant activation energy of about 200 kJ/mol. The activation energy will also be affected by the degree of branching because tertiary carbon atoms are less stable than secondary carbon atoms. Under air, the activation energy is often much lower. For example, the activation energy of PS is typically in the range of 125 kJ/mol and that of PE and PP in the range of 80 kJ/mol to 90 kJ/mol.

Chapter 7 Degradation and stabilization of different class of commodity polymers

The practical degradation process of classical polymers may be different, therefore different strategies of stabilization should be applied. In this chapter, the degradation of different classes of polymers will be discussed.

7.1 Degradation and stabilization of polyolefins

From the viewpoint of organic chemistry, olefin is a kind of hydrocarbon compound with one or more non-conjugated C=C bond structures. However, in polymer science, the meaning of polymer referred to by the term "polyolefin" is limited. Usually refers to a semi-crystalline thermoplastic material formed by homo-polymerization or co-polymerization of α-olefin (ethylene or alkyl substituted ethylene such as propylene, 1-butene, isobutene, 4-methyl-1-pentene, etc.), mainly including polyethylene, polypropylene and related co-polymers. Although the variety of these polymers is limited, they play very important role in commodity plastics, and their production is the first in all plastics. In 1997, the world's production capacity of LDPE, Linear low density polyethylene(LLDPE), HDPE and PP reached 17.5 million tons, 12.8 million tons, 21.65 million tons and 27.3 million tons respectively, with a total of 79.65 million tons. Calculated by 90% of the equipment load, a total of 71.685 million tons were produced, accounting for 52.9% of the world's plastic output of 135.5 million tons in 1997.

Due to their importance in industry, degradation of polyolefin has been widely studied and discussed. Most of the problems of polyolefin degradation are not due to the inherent structure of repeated structural units, because pure hydrocarbons which are similar to industrial polyolefins are much more stable than those under processing or use conditions. Therefore, it is believed that the degradation of polyolefin is mostly caused by external impurities. These impurities include catalysts left over from the polymerization. Advanced synthetic processes use highly active catalysts, however, small number of catalysts left do not imply the high stability of the polymer, since these highly active synthetic catalysts are also likely to be highly efficient oxidation catalysts. Meanwhile, some impurities may be introduced by the polymer during storage, transportation and processing. In addition, "pure polyolefin" only contains C—C bond, so it cannot absorb light with wavelength larger than

Chapter 7 Degradation and stabilization of different class of commodity polymers

190 nm. In the solar radiation spectrum, only light with a wavelength greater than 280 nm can reach the earth's surface (in fact, there is rarely incident radiation lower than 285 nm in most places), but industrial polyolefins without stabilizers will soon become brittle under sunlight and lose their original physical and mechanical properties. Therefore, it can be inferred that there are substances or functional groups that can absorb sunlight in polyolefins, especially these oxygen-containing groups.

Carbonyl group is the first known oxygen-containing group, and has been widely studied because of its strong UV absorption around 300 nm. Ranby and coworkers assumed that the chemical effect of ultraviolet light on polyolefin polymers was due to the presence of carbonyls in the molecular chain of the polymer, which were generated in various uncontrollable reactions during polymer synthesis or processing. It has also been proved that in the early stage of polyolefin photo-oxidation, hydroperoxide rather than carbonyl is an important initiator of photo-degradation. These substances are formed in polymer processing and are the precursor of carbonyl compounds. Although carbonyl compounds have higher absorption than hydroperoxides in the near ultraviolet region, the quantum yield of free radicals produced by carbonyl decomposition is very low. On the contrary, hydroperoxide is a weak ultraviolet absorber, but in fact, it decomposed 100% to free radicals when excited, and some of them will diffuse very quickly.

The degradation and stability of polyolefins are actually related to the structure of polymers. The thermal stability of various bonds and groups in the polymer molecular chain is as following:

$$-CC-\overset{H}{\underset{|}{C}}->-C-(-(-))>-C-\underset{|}{\overset{|}{C}}-t-\overset{C}{\underset{t}{\overset{|}{C}}}$$

$$-\underset{F}{\overset{|}{C}}->-\underset{H}{\overset{|}{C}}->-\underset{C}{\overset{|}{C}}->-\underset{Cl}{\overset{|}{C}}-$$

Therefore, the order of thermal stability of polyolefin is polyethylene>polypropylene> polyisobutylene. The temperature ($T^{1/2}$) required for half of their mass loss after 30 min heating in vacuum is 414 ℃, 387 ℃ and 348 ℃ respectively.

In the presence of oxygen, polyolefin will undergo automatic oxidation. Under high temperature and UV irradiation, oxidation is accelerated and thermal-oxidation and photo-oxidation will occur. If the polymer is first exposed to oxygen, the degradation of thermal-oxidation and photo-oxidation is further accelerated. Auto-oxidation is also related to the structure of the polymer, for example, increased chain branching in a hydrocarbon polymer will result in faster auto-oxidation. Based on this, it can be predicted that the order of oxidation stability of these three polyolefins above the melting point is high density (linear) polyethylene> low density (branched) polyethylene > polypropylene. The behavior of polyolefin when exposed to light is strongly related to their previous thermal oxidation history and the order of photo-oxidation stability of the main polyolefins measured by embrittlement time is the same as their order of thermal oxidation stability.

7.1.1 Degradation of polyolefins

1. Polyethylene

(1) Photo degradation of polyethylene. In the absence of oxygen, pure polyethylene is quite stable toward ultraviolet light. In vacuum or nitrogen atmosphere, chain breaking, dehydrogenation, cross-linking and hydrogen release may occur after long-term short UV irradiation. The physical properties of polymer affect the photochemical process. When the thickness and crystallinity of polyethylene film decrease, the crosslinking of polyethylene chain increases. Therefore, crosslinking mainly occurs in the amorphous region.

The weather resistance of commercial polyethylene is determined by the photo oxidation reaction on the sample surface, which is due to the presence of impurities or carbonyl groups. By investigation of the UV and IR spectrum of the sample after irradiation in the presence of oxygen, it is found that UV irradiation of polyethylene film in air will cause the uptake of oxygen, and results in the formation of carbonyl, hydroxyl and alkenyl, acetone, acetaldehyde, water, carbon monoxide and carbon dioxide emissions, which may induce the increase of brittleness, the formation of crosslinking and worse mechanical properties of polymer sample.

In the initial stage of polyethylene photo oxidation, the following phenomena will be observed:

1) The increase of absorption in the region of 240 – 280 nm is due to the conjugated unsaturated compounds, which are generated according to the following reactions:

$$\sim\sim CH_2-\dot{C}H-CH_2-CH_2-CH_2\sim\sim \xrightarrow{h\nu} \sim\sim CH-\dot{C}H-CH_2-CH_2-\dot{C}H-CH_2\sim\sim + H\cdot$$

$$\sim\sim CH_2-\dot{C}H-CH_2-CH_2-\dot{C}H-CH_2\sim\sim \xrightarrow{\text{radicals migration}} \sim\sim CH_2-CH=CH-CH_2-CH_2-CH_2\sim\sim$$

$$\sim\sim CH_2-CH=CH-CH_2-CH_2-CH_2\sim\sim \xrightarrow{h\nu}\cdots\longrightarrow \sim\sim CH_2-(CH=CH)_3-CH_2-CH_2-CH_2\sim\sim$$

2) The conjugated unsaturated compound disappears upon heating, and the activation energy of such reaction is only 23.4 kJ/mol. After the disappearance of conjugated bond, hydroxyl, carboxyl, carbonyl and alkenyl can be formed at a measurable rate. The concentrations of hydroxyl and carbonyl increased linearly with the increase of UV irradiation time, with activation energy for these two processes of 46 kJ/mol and 79.4 kJ/mol, respectively. The carbonyl content increases linearly with the square of irradiation time, and the activation energy of this process is 96.1 kJ/mol.

There are 1,790 cm^{-1} (γ-lactone), 1,725 cm^{-1} (carbonyl), 1,715 cm^{-1} (carboxylic acid), 1,645 cm^{-1} (alkenyl) absorption on the infrared spectrum of polyethylene after photo

Chapter 7 Degradation and stabilization of different class of commodity polymers

irradiation. The concentration of carbonyl is significantly higher in the sample with thinner thickness and higher crosslinked density. Some crosslinking bonds may contain oxygen bridges. There is broad absorption band at 1,250 – 1,170 cm^{-1} for polyethylene after ultraviolet light irradiation, indicating the presence of ether bonds.

Yanai and coworkers studied the accelerated weathering degradation of multilayer LDPE sheets. The sheet is made of 9 layers of thin films by hot pressing. It is put in an aging box at 55 ℃ for different aging times, and then the carbonyl content is measured by infrared spectroscopy. The results show that the carbonyl content is parabolic symmetrically distributed. The innermost film is least affected, and the outermost layer facing the UV irradiation source is slightly more affected, while the mechanical properties also show the same distribution. The retardation of oxygen diffusion by one or two layers of polymer on the surface greatly reduces the photo-oxidation of the inner polymer. This result shows that the photo-oxidation of polymers is autocatalytic and affected by oxygen diffusion.

(2) Thermal degradation of polyethylene. Polyethylene is also very stable to heat in the absence of oxygen. The thermal degradation of polyethylene follows the mechanism of random chain breaking and intra/intermolecular transfer reactions. The main reactions are shown in Fig. 7.1, the degradation products include ethylene, propylene, butene and olefin fragments with longer molecular chains.

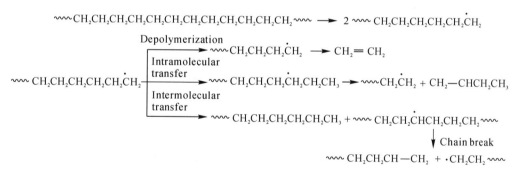

Fig.7.1 Thermal degradation of PE

Thermogravimetry and differential scanning calorimetry are effective tools to study thermal degradation. In nitrogen atmosphere, LDPE begins to decompose at about 360 ℃, and the decomposition rate reaches the highest at 430 ℃. HDPE has slightly better heat resistance, which starts to decompose at about 380 ℃, and the peak decomposition rate is about 440 ℃. Oxygen has a very important effect on thermal decomposition. The time of polyethylene loses 50% weight at 430 ℃ in nitrogen (26.7 kPa) is 20 min, while in the mixed atmosphere of nitrogen (22.1 kPa) and oxygen (5.3 kPa), it decreases to 31 s.

Heat also has an important effect on oxidation. When the temperature is high, the oxidation rate is fast, and when the temperature is low, the oxidation rate is slow.

Therefore, it is very important to avoid the contact between melted polymer and oxygen during processing. For example, when polypropylene fiber is stretched to 300% at the rate of 20%/min, the peroxide content increases by 10 times compared with that before stretching due to their contact with oxygen during the stretching process.

After thermal oxidation of polyethylene, ketones, carboxylic acids and volatiles are formed. There are also a small amount of esters and γ-lactones. When oxidized at high temperature of 264 - 289 ℃, there are also aldehydes in the volatile. During the thermal oxidation of polyethylene, the chain fracture is not the main reaction, but the main reaction is the crosslinking reaction or the formation of long-chain branched products. It is generally believed that crosslinking is formed by the combination of free radicals (including alkyl radicals, alkoxy radicals and peroxyl radicals). But there is evidence that the addition reaction of free radicals on double bonds is very important. It was found that the melt flow rate of hydrogenated HDPE did not decrease after heat treatment. In other experiments, it was found that the amount of double bond decreased with the increase of crosslinking degree or molecular weight. The study of linear low density polyethylene indicated that in the torque rheometer test, the increase of torque was related to the rapid decrease of vinyl groups, but the concentration of vinylidene and trans vinylidene groups did not change. In addition, there was no crosslinking of hydrogenated linear low density polyethylene alkyl radicals, alkoxy radicals and peroxyl radicals may react with vinyl to cause crosslinking (see Fig. 7. 2).

$$R\cdot + H_2C=CH-CH_2\sim \rightarrow R-CH_2-\overset{\cdot}{C}H-CH_2\sim$$
$$RO\cdot + H_2C=CH-CH_2\sim \rightarrow R-O-CH_2-\overset{\cdot}{C}H-CH_2\sim$$
$$RO_2\cdot + H_2C=CH-CH_2\sim \rightarrow R-O-O-CH_2-\overset{\cdot}{C}H-CH_2\sim$$

Fig.7.2 Crosslink reaction between radical and vinyl groups

(3) Study on the application of polyethylene degradation. Pages and others studied the structural changes and mechanical properties of HDPE aged by climate in winter in Canada with FTIR and DSC. It is found that the results of the two research methods are the same. After aging, the crystallinity of the polymer decreases, which greatly reduces the impact properties of the material, but other properties are not greatly affected.

Svorcik and others studied the surface modification of polyethylene and polypropylene by positive ion treatment. After ion treatment, the polymer macromolecules broke their chains, and it was observed that oxygen penetrated into the areas damaged by radiation, and therefore polyethylene was easier to be oxidized. While polypropylene shows higher surface polarity and conductivity than polyethylene. After electron beam irradiation (14.89 MeV, 247 Gy/min) of polyethylene, the surface polarity of the sample increased linearly with the

Chapter 7 Degradation and stabilization of different class of commodity polymers

irradiation dose, but its crystallinity and melting temperature did not change significantly.

2. Polypropylene

(1) Photo degradation of polypropylene. Pure polypropylene does not absorb light greater than 200 nm. Chromophores such as catalyst residue, hydroperoxide group, carbonyl group and double bond may exist in industrial grade polypropylene.

Some chromophores in polypropylene are listed in Table 7.1.

Table 7.1 Some chromophores in polypropylene and formation rate of radical

Chromophore	Approximate concentration/(mol·L)	Formation rate of radical/(mol·s)
Hydroperoxide	$10^{-4} - 5 \times 10^{-3}$	$\approx 5 \times 10^{-3}$
Kelone	$< 10^{-3}$	$< 10^{-8} - 10^{9}$
Residual catalyst [such as $TiOCl_2$, $Ti(OBu)_a$]	$10^{-4} - 10^{-3}$	$\approx 6 \times 10^{-8}$
Charge transfer complex (such as $P \cdots O_2$)		$\approx 4 \times 10^{-11}$

All of these groups absorb ultraviolet light with a wavelength greater than 290 nm. Industrial polypropylene resins may also contain antioxidants and peroxide decomposing agents, which also extend the absorption range of polymers to wavelength greater than 200 nm. Methyl radicals, methylene radicals and chain end radicals were formed when polypropylene was irradiated with ultraviolet light at 77 K in vacuum.

When polypropylene is irradiated with ultraviolet light in the air, it will undergo rapid photo-oxidation, and its mechanical properties and other physical and mechanical properties will change significantly. The molecular weight of polypropylene decreases rapidly with the extension of UV irradiation time. The number of chain breaks initially has a linear relationship with the irradiation time. Subsequently, the chain breaking rate increased with irradiation time, showing an auto-catalytic photo-oxidation mechanism. After a long term of irradiation, the polypropylene film generates fine cracks and becomes brittle. The infrared spectrum study shows that after irradiating polypropylene with ultraviolet light, the following absorption bands are generated: hydroxyl and hydrogen bonded hydroperoxide (3,400 cm^{-1}); carbonyl (1,715 - 1,720 cm^{-1}); γ-lactone (1,728 cm^{-1}); carboxylic acid (1,715 cm^{-1}); alkenyl (1,645 cm^{-1}). There is about 16% aldehyde and ketone groups in the product of photo-oxidative degradation of polypropylene, while in the case of thermal oxidative degradation of polypropylene it contains 42%. Oxidation occurs preferentially on the surface of polypropylene film, but also in the interior of the sample to a certain extent. With the increase of the effective depth of light penetration, the concentration of hydroperoxide and carbonyl decreases.

Hydroperoxide is also the main initiating photo-initiator for the photo-oxidation of polypropylene. The kinetic chain length of photo-oxidation of polypropylene is about 10 times that of polyethylene, so the concentration of hydroperoxide in photo-oxidized polypropylene is much higher than that in oxidized polyethylene. Under UV irradiation, tertiary hydroperoxide is decomposed into tertiary alkoxy and hydroxyl radicals, and then tertiary alkoxy radicals occur β-fracture reaction. As shown in Fig. 7.3, reactions (b) and (c), β-fracture reaction is shown in Fig. 7.3 (b), which may be the main cause of the main chain fracture. Electron spin resonance (ESR) studies showed that methyl radicals were generated at the same time as ketone products. It is reported that about 15% of the hydrogen peroxide groups that have suffered photolysis undergo β-fracture reaction.

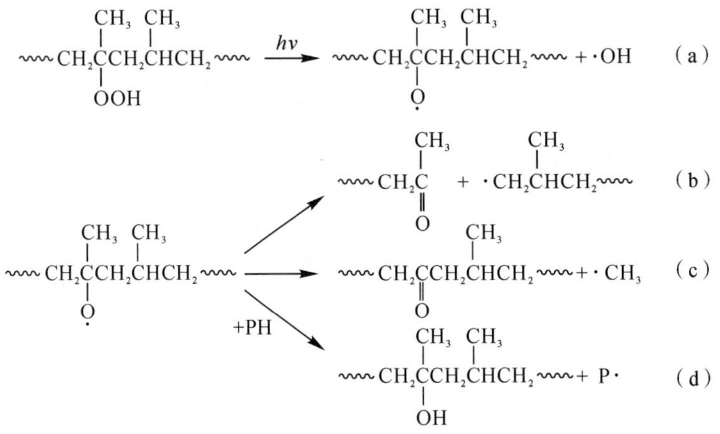

Fig.7.3 Photo degradation of PP

In the process of photolysis of polypropylene hydroperoxide, the main volatile product is water, which is generated by the reaction between hydroxyl radicals and proton. Other products are a small amount of ethane, ethylene, propane and propylene.

Tidjani studied the effect of the power of γ-ray on the oxidation and physical properties of polypropylene. In the power range of 0.02 – 1 Gy/s (2 – 100 rad/s), the higher the power (the same total dose), the smaller the change of elongation at break of the sample. At the same time, it is found that the change of elongation at break is related to the change of relative molecular weight. Under γ-ray irradiation, the oxidation mechanism of polypropylene to ketone and ester is different. Martakis found that isotactic polypropylene with a wide relative molecular weight distribution and smaller molecular weight shows higher γ-ray stability.

(2) Thermal degradation of polypropylene. Polypropylene is 3.5 times more sensitive to thermal cracking in nitrogen than polyethylene, and about 30 times more sensitive to thermal cracking in the mixture of nitrogen (21.2 kPa) and oxygen (5.3 kPa). When thermally degraded in vacuum, polypropylene starts to produce volatiles at a temperature 100 ℃ lower than polyethylene. The thermal decomposition of polypropylene also follows the random chain breaking mechanism, generating primary and secondary radicals. Due to the intra/

Chapter 7 Degradation and stabilization of different class of commodity polymers

intermolecular radical transfer to tertiary carbon atoms and β-fracture, secondary free radicals are generated, resulting more secondary free radicals than primary free radicals (Fig. 7.4). The products of thermal decomposition mainly include propylene, 2-pentene, 2,4-dimethyl-1-heptene, as well as methane, ethane, propane, etc. The main volatile products (in order of importance) of random polypropylene are 2,4-dimethyl-1-heptene, 2-pentene, propylene, 2-methyl-1-pentene and a small amount of isobutene, the product order is not affected by isotacticity and temperature (383 - 438 ℃). In the degradation mechanism of polypropylene shown in Fig. 7.4, the decrease of the molecular weight of the polymer is mainly attributed to random chain breaking and intermolecular chain transfer of primary radical (B), including the reaction of capturing tertiary hydrogen atoms from other polymer molecules. The production of volatiles is mainly attributed to the depolymerization of tertiary radical (A) and intramolecular chain transfer reaction.

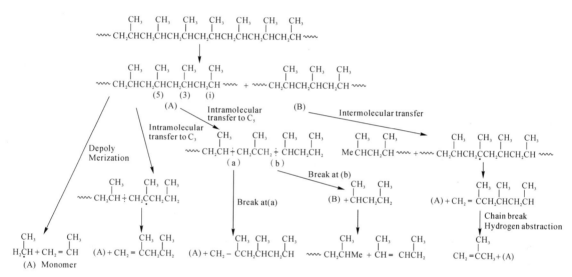

Fig.7.4 Mechanism for the thermal degradation of PP

In the presence of oxygen, the reaction of free radicals will inevitably lead to the formation of hydroperoxide. There is a lot of evidence that polyolefins generally contain hydroperoxide groups after processing. Such group is more sensitive to heat or the interaction between heat and transition metals. Therefore, once the hydroperoxide group is formed, it will greatly promote the degradation of polymers. After thermal oxidation of polypropylene, aldehydes, ketones, carboxylic acids, esters and γ-lactone and other structures and the volatile products during thermal oxidative degradation of polypropylene (in order of importance) include water, formaldehyde, acetaldehyde, acetone, methanol, hydrogen peroxide, carbon monoxide and carbon dioxide. The changes of physical properties are mainly manifested in the decrease of molecular weight and the resulting deterioration of mechanical properties.

Macromolecular alkoxy radical is a key intermediate in the thermal oxidation of polypropylene. It is produced by the decomposition of macromolecular hydroperoxide, or by the non terminating reaction of two tertiary peroxy radicals:

$$2\sim\sim CH_2\underset{\underset{OO\cdot}{|}}{\overset{\overset{CH_3}{|}}{C}}CH_2\overset{\overset{CH_3}{|}}{C}H\sim\sim \longrightarrow 2\sim\sim CH_2\underset{\underset{O}{|}}{\overset{\overset{CH_3}{|}}{C}}CH_2\overset{\overset{CH_3}{|}}{C}H\sim\sim + O_2$$

It is generally believed that chain breaking is caused by the β-fracture reaction of macromolecular alkoxy radicals (as shown in reaction (b) in Fig. 7.3). It was previously believed that, the termination reaction occurs immediately after β-fracture. However, Carlson found that in the isotactic polypropylene irradiated by γ-ray, the disappearance of each alkoxy radical will lead to 4-5 times β-fracture. As revealed by ESR study, peroxyl radicals were also found in the crystalline region of polypropylene, indicating that oxygen can also diffuse into the crystalline region and generate oxidation reaction.

Kim and coworkers simulated the thermal degradation and peroxide induced degradation of polypropylene in a twin-screw extruder by using computer. The effects of screw form, screw speed, processing temperature and initiator dosage on degradation were studied, and the simulation results were compared with the experimental results. It is found that different screw forms have a great impact on the degradation of polypropylene (see Fig. 7.5). In Fig. 7.5, screw A has no kneading element, and screws B and C have only one kneading element, but the kneading element of screw C is closer to the hopper than screw B, and screw D has three kneading elements. Kneading elements have an important impact on melting, mixing, heating and residence time. It can be seen from the figure that it also increases the degradation of polymers. Fig. 7.6 shows the comparison between the test results and the simulation results. The higher the temperature is, the more serious the degradation of polypropylene is.

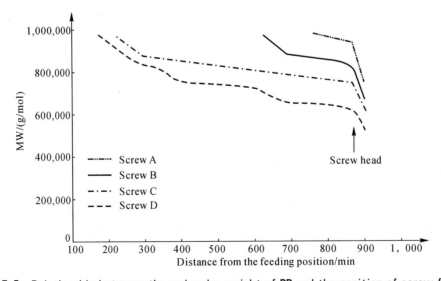

Fig.7.5 Relationship between the molecular weight of PP and the position of screw froms

Chapter 7　Degradation and stabilization of different class of commodity polymers

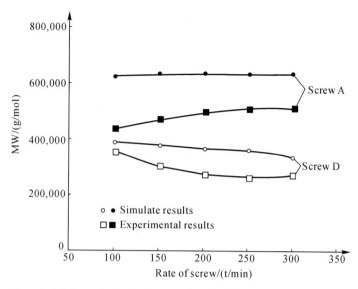

Fig.7.6　Experimental data and simulation results for PP under different condition of screw

3. Ethylene-propylene copolymer and polyethylene-polypropylene blend

When the ethylene-propylene copolymer was irradiated with 365 nm ultraviolet light, no ketone was formed on the molecular chain. On the contrary, the ketones on the original molecular chain were also reduced. Meanwhile, the formation of carboxylic acids, esters and peroxides were observed. It was previously believed that alkoxy radicals and hydroxyl radicals formed by the decomposition of hydroperoxide reacted with ketones to form esters and carboxylic acids, but now it tends to assume that the complex of hydroperoxide and ketone generates carboxylic acids and ethers by the interaction with light.

Sarwade and coworkers studied the phase structure of heterogeneous ethylene-propylene copolymer during photo-oxidation by SEM, and compared the difference between isotactic polypropylene and linear low density polyethylene. Under the irradiation of polychromatic ultraviolet light with wavelength larger than 290 nm in the air at 55 ℃, the surface of the sample is damaged. The degree of surface damage of the sample has a good correlation with the photo-oxidation of the polymer. The photo-oxidation of ethylene-propylene copolymer is much slower than that of isotactic polypropylene.

Polyolefins are often modified by grafting polar compounds (such as maleic anhydride) initiated by peroxides. However, there is usually complicated side reactions for macromolecular free radicals. Generally, polypropylene tends to undergo β-fracture to degradation, while polyethylene tends to crosslink. Whether these two tendencies can achieve a certain balance in PP/PE blends has become a hot topic of concern. Yu and coworkers studied peroxide and irradiation chemical modification of LDPE/PP alloys. Two different peroxide concentrations (mass fraction 0.1% and 1%) and two irradiation doses (30 kGy and 100 kGy) were studied experimentally. The modification effect was characterized by rheological, mechanical properties and microstructure observation. The results showed that polyethylene underwent crosslinked and polypropylene degraded under these conditions. In the presence of peroxide, the degradation of PP exceeded the crosslinking of PE, but under the presence of irradiation, the crosslinking of PE exceeded the degradation of PP. For the alloy with PP as the main component, irradiation shows better improvement of the properties of the alloy than peroxide. For the alloy with the same composition of PP as the main component, the viscosity and elasticity of the melt under irradiation are higher than that under peroxide.

Braun and coworkers have studied the degradation, crosslinking and grafting reactions of PP/PE alloy initiated by peroxide in melt state and solution state. According to the observation results of viscosity, GPC and DSC, the reactions of PP and PE in the melt of PP/PE alloy are obviously independent of each other. In other words, PP degraded and PE underwent crosslink, and there is no graft polymer of PP and PE. The graft polymer of PP and PE appears in the solution of PP/PE alloy. The above results show that although the PP/PE alloy melt is transparent, it is still a two-phase system. While the solution of PP/PE alloy may be a homogeneous single-phase system, in which degradation and crosslinking reactions are suppressed.

4. Polyisobutylene

Each repeating unit of polyisobutylene contains a quaternary substituted carbon atom. This structure improves the photostability in comparison with polypropylene. The first article on the UV degradation of polyisobutylene in vacuum was published by Garstensen et al. Based on ESR data, they proposed the following mechanism: ①the C—C bond on the main chain breaks, producing two free radicals. ②the C—C bond on the side radical breaks, producing another two free radicals. ③ the generated free radicals can attack the polymer molecules and abstract hydrogen atoms from the methylene group on the main chain, as shown in Fig. 7.7 (c) or capture hydrogen atom from side methyl group, as shown in Fig. 7.7 (d); ④disproportionation reaction is the main reason for the decrease of molecular weight, as shown in Fig. 7.7 reaction (e)(f)(g). A new double bond will be generated after each break of the main chain breaks from these reactions.

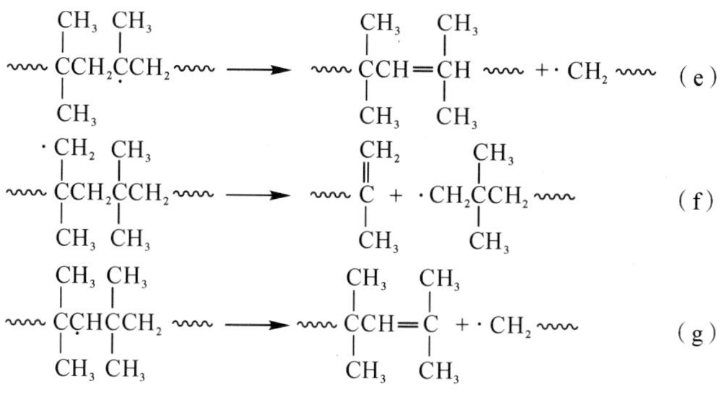

Chapter 7 Degradation and stabilization of different class of commodity polymers

$$\sim\sim\underset{\underset{CH_3}{|}}{\overset{\overset{CH_3}{|}}{C}}CH_2\underset{\underset{}{|}}{\overset{\overset{CH_3}{|}}{C}}CH_2\sim\sim \longrightarrow \sim\sim\underset{\underset{CH_3}{|}}{\overset{\overset{CH_3}{|}}{C}}CH=\underset{\underset{CH_3}{|}}{\overset{\overset{CH_3}{|}}{C}}H\sim\sim + \cdot CH_2\sim\sim \quad (e)$$

$$\sim\sim\underset{\underset{CH_3}{|}}{\overset{\overset{\cdot CH_2}{|}}{C}}CH_2\underset{\underset{CH_3}{|}}{\overset{\overset{CH_3}{|}}{C}}CH_2\sim\sim \longrightarrow \sim\sim\underset{\underset{CH_3}{|}}{\overset{\overset{CH_2}{\parallel}}{C}} + \cdot CH_2\underset{\underset{CH_3}{|}}{\overset{\overset{CH_3}{|}}{C}}CH_2\sim\sim \quad (f)$$

$$\sim\sim\underset{\underset{CH_3}{|}}{\overset{\overset{CH_3}{|}}{C}}CHC\underset{\underset{CH_3}{|}}{\overset{\overset{CH_3}{|}}{C}}CH_2\sim\sim \longrightarrow \sim\sim\underset{\underset{CH_3}{|}}{\overset{\overset{CH_3}{|}}{C}}CH=\underset{\underset{CH_3}{|}}{\overset{\overset{CH_3}{|}}{C}} + \cdot CH_2\sim\sim \quad (g)$$

Fig.7.7 Photo degradation of polyisobutylene

7.1.2 Stabilization of polyolefins

The industrial polyolefin without stabilizer will degrade violently in the sun, and soon become brittle and lose its original physical and mechanical properties. At the temperature below the melting point, the degradation mainly occurs in the amorphous region. The chain breaking in the random phase leads to the readjustment of the polymer chain and improves the crystallinity. Finally, surface cracks are generated. The stress concentration of crystal discontinuous bonds and surface cracks can promote crack growth and lead to polymer embrittlement.

Oxygen plays the most important role in the degradation of polyolefins, and the stability of polyolefins is mainly attributed to antioxidant degradation. Oxidation is a chain reaction, and the carrier of the chain is alkyl and peroxyalkyl radicals, whose relative concentration depends on temperature and oxygen pressure. At a typical processing temperature, there is very small concentrations of alkyl radicals in any polyolefin. At high temperature, even a small amount of oxygen will have a great catalytic effect on the degradation of molten polymer. The reaction proceeds in the following order:

$$R\cdot + O_2 \rightarrow R\cdot O_2 \quad (7-1)$$

$$RO_2\cdot + RH \rightarrow RO_2H + R\cdot \quad (7-2)$$

$$RO_2H \rightarrow RO\cdot + \cdot OH \quad (7-3)$$

$RO\cdot$ and $\cdot OH$ radicals can abstract hydrogen from the polymer to form alkyl radicals. Reaction (7-3) rapidly increases the total concentration of free radicals in the polymer, and the degradation process is automatically accelerated. At the same time, a considerable proportion of alkoxy radicals undergo β-chain-fracture to produce a carbonyl group and an alkyl terminal radical.

In principle, compounds that can capture alkyl radicals and peroxide alkyl radicals can interrupt the above chain reaction. Many free radical capture radicals are effective melt stabilizers.

The relative rate of reaction formula (7-1) and reaction formula (7-2) basically determines the relative concentration of alkyl radicals and peroxyalkyl radicals in the polymer melt. Reaction formula (7-1) is a first-order reaction of oxygen concentration and is controlled by collision, so its rate constant is independent of temperature. Reaction formula (7-2) has considerable activation energy, and its reaction rate is independent of oxygen concentration.

1. Stability of melt polymer

Under processing conditions, due to the low solubility of oxygen in the polymer at processing temperature and the limited air in contact with the polymer, the concentration of oxygen is very low. Radical captures with high reactivity with alkyl radicals are effective melt stabilizers. The main compounds are quinones, which are also familiar polymerization inhibitors, such as benzoquinone and p-benzoquinone, and quinone compounds derived from hindered phenols. Their stabilization mechanism is mainly due to their effective oxidation, which can combine with alkyl radicals (get an electron) to become stable radicals:

$$O=\bigcirc=O + R\cdot \longrightarrow RO-\bigcirc-O\cdot$$

The generated stable radicals can also take a hydrogen atom from the alkyl radicals (disproportionation termination) and become phenol. If it is hindered phenol, it can give hydrogen atoms to become stable free radicals. In this way, one molecule of stabilizer can act multiple times. Therefore, one in ten thousand (mass) of hindered phenol (or quinone) can effectively stabilize PE and PP melts.

$$O=\bigcirc_{tBu}^{tBu}-CH-\bigcirc_{tBu}^{tBu}-O\cdot \underset{-e(-H)}{\overset{+e(+H)}{\rightleftarrows}} O=\bigcirc_{tBu}^{tBu}=CH-\bigcirc_{tBu}^{tBu}-OH$$

Under the condition of oxygen enriched at low temperature, the alkyl peroxide radical in the polymer is dominant, so the alkyl peroxide radical scavenger is a more effective stabilizer. The most common formulations are those containing 2,6-di-$tert$-butylphenol, which can give the hydrogen of hydroxyl group to the peroxyl radical to form a very stable radical, so that oxidation cannot be reinitiated. Such free radical is an effective anti-thermal oxidant. Its main disadvantage is that it is easy to produce colored oxidation products, and their photo-degradation stability is poor, which limits their use as UV stabilizers.

Another important reaction leading to polymer chain breaking during PP processing is reaction formula (7-3). Therefore, the reagent that decomposes hydroperoxide into alcohol according to the non radical process is a very effective melt stabilizer. The two main types of antioxidants that decompose hydroperoxide are phosphite and various sulfides, such as catechol phosphite. In addition, thermal and UV stabilizers for some polyolefins, such as dithiocarbamates of transition metals and similar dithiophosphates, as well as sulfonates. In general, soluble zinc complexes are more effective. The characteristics of the mechanism of decomposition of peroxides by metal dithiol salts is shown in Fig. 7.8.

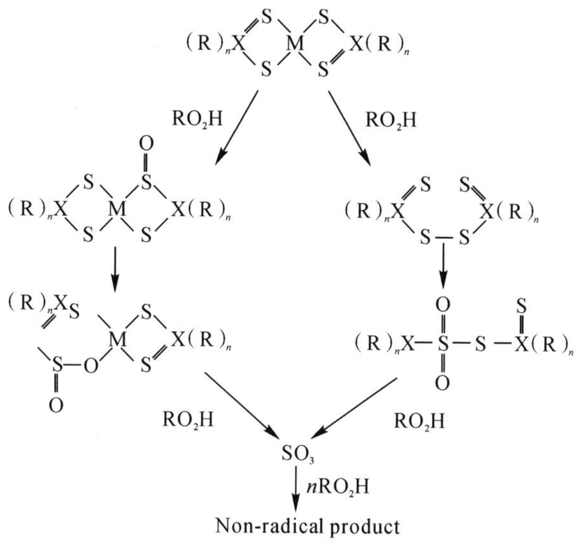

Chapter 7 Degradation and stabilization of different class of commodity polymers

Fig. 7.8 Mechanism for the decomposition of peroxides by transition metals dithiocarbamates complex

2. Stabilization of polyolefin products

The stabilization of polyolefin products used at low temperature is mainly photo oxygen stabilization, while the thermal oxidation stability of products used at high temperature must also be considered.

(1) Stabilization of polyolefins at high temperature. There are many fiber reinforced polypropylene components used in high temperature, such as engine parts, air distributors, etc. Air oven test can be used to check the stability of polypropylene, or as a preliminary "screening test" of antioxidant activity. At 140 ℃, the un-protected polypropylene oxidizes very quickly, and the polymer becomes brittle within 1 h. Even at 120 ℃, the useful lifetime of the polymer is less than 10 h. Oven aging is usually carried out in a tubular multi chamber oven, with only one sample in each chamber to avoid cross transfer of antioxidants between samples. When the sample is thin (< 0.03 mm), the diffusion of oxygen is not a step to determine the rate since there will be a lot of oxygen in the sample. The most effective degradation inhibitor is hindered phenol antioxidant. The same antioxidant is more effective when the sample is thicker and the temperature decreases, and the degradation of the sample becomes slower. Under high temperature conditions, the volatility of antioxidant has an important impact on its stability. The lower the volatility, the better the effect. If the sample thickness is large, the oxygen in the centre of the sample may be exhausted at high temperature. Therefore, quinones that can capture alkyl radicals may be more effective. The thermal stability of un-stabilized polypropylene decreases with the extension of processing time. Polypropylene stabilized with 2,6-di-tert-butyl-p-methylphenol (BHT) will be oxidized to stilbene quinone during the processing of polypropylene, and the thermal stability of the product will be more

stable with the increase of the severity of processing conditions.

The lower the volatility of the antioxidant, the better the antioxidant effect. A logical development of this concept is to chemically bind the antioxidant to the polymer. A more feasible method is to graft antioxidants onto polyolefin molecules when processing polyolefins. Sulfur-containing hindered phenols and some thiol antioxidants are easy to produce high-level chemical combination through mechanochemical processes and polyolefin reactions.

In addition, hydroperoxides are ubiquitous initiators for both thermal oxidation and photo-oxidation. Therefore, the use of peroxide decomposing agents in combination with hindered phenolic antioxidants will produce synergistic effects and obtain much better antioxidant effects. The effectiveness of sulfur-containing hindered phenolic antioxidants is also much higher than that of pure hindered phenolic antioxidants. This is because, in addition to their normal chain termination, they also have the ability to catalyze the decomposition of peroxides. This effect shown by the same compound is also called self synergism.

(2) Photo stabilization of polyolefins. The photo-degradation of polyolefins is mainly manifested by photo-oxidation. There are two main interrelated oxidation cycles: firstly, the formation of polymer radicals (alkyl and alkyl peroxyl); Secondly, the generation of free radical initiators, the most important of which is hydrogen peroxide. The pyrolysis or photolysis of hydroperoxide leads to the generation of more free radicals and chromophores that further absorb light, such as carbonyl compounds. These oxidative degradation cycles can be blocked by two types of stabilizers:

1) Chain termination antioxidants. They capture free radicals (alkyl, alkylperoxyl, or both) generated during chain growth to interfere with the first degradation cycle.

2) Preventive antioxidants. They scavenge or stabilize potential free radical generators (hydroperoxides) while disturbing the second cycle.

(3) Stabilization of chain terminated antioxidants. As mentioned above, different free radical scavengers have different ability to capture alkyl and alkyl peroxyl groups. Therefore, when selecting antioxidant, the ratio of $[R\cdot]$ and $[ROO\cdot]$ under specific conditions shall be considered, and this ratio is a function of oxygen concentration. In photo-oxidation, the initiation rate is higher than the rate of oxygen diffusion into the polymer, which makes the ratio of $[R\cdot]/[ROO\cdot]$ greater than two orders of magnitude in liquid hydrocarbons. Although macromolecular alkyl radicals are easily oxidized by a variety of oxidants (electron acceptors), few of them are light stable (quinone, nitrone, phenoxy radicals). Therefore, even those containing quinone groups (such as stilbene quinone and peroxide diketene) can capture alkyl radicals, they are still photosensitizers in essence. Hindered phenol is also an effective heat stabilizer. As an electron donor, it can give hydrogen atom to terminate alkyl peroxyl radical. Due to the instability of phenol and its oxidation products to ultraviolet light, it has only limited value as a light stabilizer for

Chapter 7 Degradation and stabilization of different class of commodity polymers

polyolefins, but they can effectively synergize with ultraviolet absorbers. The primary role of the latter may be to prevent the photolysis of the former. Some UV stabilizers combine these two functions in the same molecule.

Stabilizers that can capture only one of [R·] and [ROO·] usually have 1-2 stoichiometric inhibition coefficients (the number of kinetic chains that can be blocked by each mole of antioxidant), and the inhibition coefficients of chain termination antioxidants that can capture both [R·] and [ROO·] are much greater than one. Therefore, antioxidants with the ability to alternate between oxidation and reduction states can show a catalytic regeneration mechanism and thus have great potential for stabilizing polymers. Stable nitroso radicals belong to this category. Hindered aliphatic amines have a very good role in protecting polyolefins from UV light, and its effectiveness is derived from its conversion product-nitroso radical. Nitroso radicals and hydroxylamine produced by nitroso radicals are effective UV stabilizers. They also play the same role in the stabilization of PP melt. The high efficiency is due to the complementarity of donor-acceptor-antioxidant mechanism. Stabilization mechanism of nitroso radical as ultraviolet stabilizer of polypropylene and important redox reaction (see Fig. 7.9).

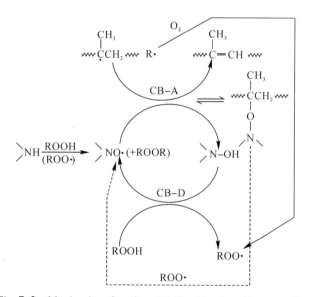

Fig.7.9 Mechanism for the stabilization by nitroso radical

(4) Stabilization of preventive stabilizers. Preventive stabilizer reduces chain initiation rate. They inhibit the photodegradation rate caused by physical and chemical processes and act as UV stabilizers for polyolefins. preventive stabilizers include three categories: ① prevent ultraviolet rays from polymers. ② excited quenching agent. ③ light stabilizer for decomposing hydroperoxide.

1) UV shielding agent and UV absorber. Compounds that are opaque to light or have strong spectral absorption in the range of 300-400 nm can act as ultraviolet absorbers as long

as they have some ways to dissipate energy innocuously (e.g. through non radiation processes). The deposition of reflective or opaque pigments on the surface can protect the polymer and play a role in shielding ultraviolet rays. Ultraviolet absorber also plays a role in shielding ultraviolet rays, but the contribution of shielding to the overall mechanism of stabilization is very small, which can be proved by the following tests: the embrittlement time of the unstable PP film is much shorter than that of the previous stabilized film when the unstable PP film is directly irradiated behind the stabilized film without contact. The effect of ultraviolet absorber is weakened with the increasing severity of polymer processing conditions.

2) Excited state quencher. The energy transfer from the photoexciter [S] to a quencher molecule [Q] can dissipate the obtained energy innocuously, but effective energy transfer can be realized only when the triplet energy of the quencher (stabilizer) is lower than the chromophore. When the emission spectrum of S and the absorption spectrum of Q overlap greatly, long-range energy transfer [S] to [Q] can occur. This can be an effective way of energy transfer when the quencher concentration is low, whereas the energy transfer according to the collision mechanism becomes important only when the quencher concentration is high.

3) Light stabilizer for decomposing hydroperoxide. The hydroperoxide is decomposed into stable compounds in the way of non radicals.

7.2 Degradation and stabilization of polyvinyl chloride and other chloride-containing polymers

The thermal stability of polyvinyl chloride (PVC) is very poor, and the initial decompositions temperature is far lower than its viscous flow temperature. PVC cannot be processed into products by melting without heating stabilizer. Under high temperature, PVC will quickly turn black and brittle. PVC used to occupy the largest position of plastic production for a long time. Its applications and developments are inseparable from its degradation and stability research. Therefore, there are many studies on the degradation and stability of PVC.

7.2.1 Degradation of polyvinyl chloride

1. Thermal degradation of polyvinyl chloride

(1) Characteristics of PVC degradation. The typical characteristic of PVC degradation is the release of HCl. The degradation in oxygen is much faster than that in inert gas and vacuum. Even the presence of trace oxygen will greatly accelerate the decomposition of PVC. For example, the comparative study on the release rate of HCl from polyvinyl chloride in high-purity nitrogen and industrial nitrogen confirms that although the industrial nitrogen contains only a small amount of oxygen, it is sufficient to significantly accelerate the rate of

Chapter 7 Degradation and stabilization of different class of commodity polymers

HCl removal reaction. Under normal circumstances, oxygen cannot be completely eliminated. As long as there is oxygen, PVC will undergo oxidation reaction. Therefore, the actual degradation reaction of PVC is complicated, several chemical reaction processes are often carried out at the same time, including decomposition and HCl removal, oxidative chain breaking and cross-linking, and a small amount of aromatization.

It is generally believed that the HCl released from the decomposition of polyvinyl chloride has a catalytic effect on its further removal of HCl (so called autocatalysis). For example, the thick sample degrades faster than the thin sample (see Fig. 7.10), which is due to the catalytic role of the slowly released HCl in the thick sample.

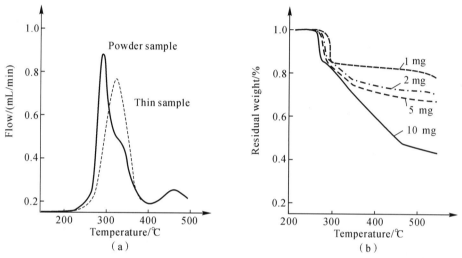

Fig.7.10 Effect of the thickness on the degradation of polymer:
(a) the degradation of PVC in vacuum; (b) the degradation of PAN in N_2

The thermogravimetric study shows (see Fig. 7.11) that the decomposition of PVC is carried out in two stages under the temperature programmed condition. The decomposition of the first stage is related to the removal of HCl, and there may be a small amount of benzene and ethylene. Most of the degradation products in the second stage are dark colored non-volatile cyclic fragments, and the composition of volatile components in this stage is also very complicate, including aromatic and alicyclic compounds, and some carbonized residue left at 500 ℃.

The color gradually turns red when HCl is released from the degradation of polyvinyl chloride, which is attributed to the formation of polyene structure, which can be proved by the UV spectrum (see Fig. 7.12). But the average length of the polyene structure is not long, about 6-10 conjugated double bonds. During the release of HCl, the molecular weight of the polymer increases due to crosslink and cyclization.

(2) Mechanism of removal of HCl. Why is the thermal stability of polyvinyl chloride much worse than that of the model compound, and HCl is released at a temperature much

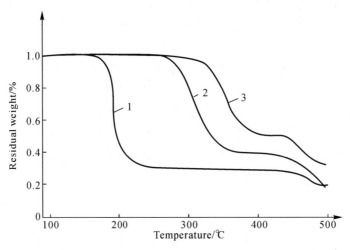

Fig.7.11 TGA traces of the halogen-containing polymer under N_2 with heating rate of 10 ℃/min: 1 for polyvinyl bromide, 2 for polyvinyl chloride, 3 for polychloroprene

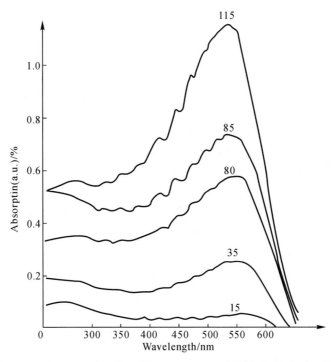

Fig.7.12 UV-vis spectra for the different stage of PVC degradation in solution (150 ℃, N_2), which was prepared via suspension

lower than that of the 1,3-chloro model compound? It is believed that the abnormal structures on the polymer molecular chain, such as branching, chloroallyl group, oxygen-containing structure, end group, head-to-head structure, trigger the dehydrochlorination reaction. The most influential point is that β-chloroallyl group (—CH=CHCHCl—) is the

Chapter 7 Degradation and stabilization of different class of commodity polymers

most important to initiate dehydrochlorination reaction, which can be inferred from the stability order of a series of chlorine containing model compounds (see Fig. 7.13).

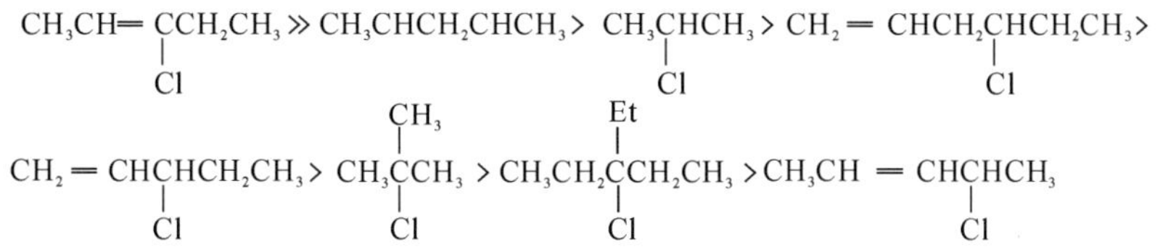

$$\underset{\underset{Cl}{|}}{CH_3CH=CCH_2CH_3} \gg \underset{\underset{Cl}{|}}{CH_3CHCH_2CHCH_3} > \underset{\underset{Cl}{|}}{CH_3CHCH_3} > CH_2=\underset{\underset{Cl}{|}}{CHCH_2CHCH_2CH_3} >$$

$$CH_2=\underset{\underset{Cl}{|}}{CHCHCH_2CH_3} > \underset{\underset{Cl}{|}}{CH_3\overset{\overset{CH_3}{|}}{C}CH_3} > \underset{\underset{Cl}{|}}{CH_3CH_2\overset{\overset{Et}{|}}{C}CH_2CH_3} > CH_3CH=\underset{\underset{Cl}{|}}{CHCHCH_3}$$

Fig.7.13 Stability sequence for some typical chlorine-containing compounds

On the other hand, Minsker and coworkers believed that carbonyl allyl group (—COCH=CHCHClCH$_2$—) is the most important initiation point of dehydrochlorination reaction. According to the stability order of the model compound, the stability of PVC (with high branching degree) polymerized at higher temperature is decreased, indicating that branching is also unfavorable to the stability of PVC. However, there is no obvious difference between the stability of PVC with head-to-head structure and that of PVC with normal structure.

As for the mechanism of decomposition and dehydrochlorination of polyvinyl chloride, there are mainly free radical mechanism, ion and ion-molecule mechanism, and simultaneous (but not combined) free radical and ion-molecule mechanism.

1) Free radical mechanism. It has been confirmed that there are indeed free radicals in the degradation process of polyvinyl chloride, and the most convincing evidence is: ① chain transfer reaction of PVC degradation in the presence of labeled toluene. ② in the mixture of PVC and PMMA, the free radical depolymerization reaction of PMMA occurred due to the decomposition of PVC at a temperature far lower than the normal degradation temperature of pure PMMA, that is, at the dehydrochlorination temperature of PVC (225℃). ③ ESR studies showed that there were free radicals in the degradation of PVC. ④ azodiisobutyronitrile can initiate dehydrochlorination of PVC.

The free radical chain mechanism includes randomly initiating the C—Cl bond breaking reaction at a position with poor stability, generating macromolecular free radicals, and then directly removing HCl or generating chlorine atoms in advance to dehydrochlorination. The main modes of chain initiation, chain growth and chain termination reactions are shown in Fig. 7.14.

However, the free radical mechanism of dehydrochlorination of polyvinyl chloride cannot explain the autocatalytic effect of HCl, nor the catalytic effect of acetic acid and Lewis acid on dehydrochlorination.

2) Ion-molecular mechanism. According to this mechanism, the initiation of decomposition of polyvinyl chloride to remove HCl is caused by the C—Cl polar bond and the adjacent C—H bond activated by its energy, resulting in the formation of a tetracyclic ion complex. When the tetracyclic ion complex decomposes, HCl escapes and forms a double

bond in the polyvinyl chloride molecule (see Fig. 7.15).

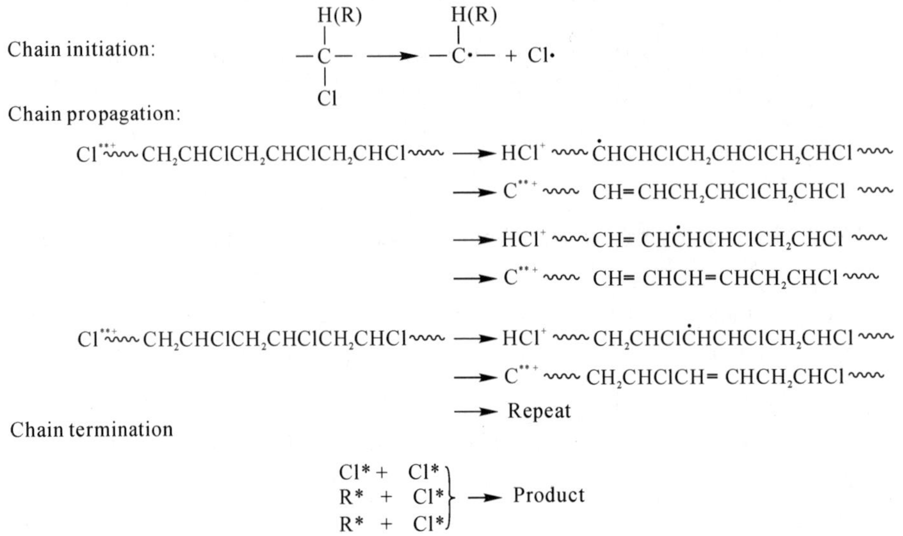

Chain initiation:

$$-\underset{\underset{Cl}{|}}{\overset{\overset{H(R)}{|}}{C}}- \longrightarrow -\overset{\overset{H(R)}{|}}{C}\cdot - + Cl\cdot$$

Chain propagation:

Cl·∼∼ CH₂CHClCH₂CHClCH₂CHCl ∼∼ ⟶ HCl⁺ ∼∼ ĊHCHClCH₂CHClCH₂CHCl ∼∼

⟶ C··⁺ ∼∼ CH=CHCH₂CHClCH₂CHCl ∼∼

⟶ HCl⁺ ∼∼ CH=CHĊHCHClCH₂CHCl ∼∼

⟶ C··⁺ ∼∼ CH=CHCH=CHCH₂CHCl ∼∼

Cl·∼∼ CH₂CHClCH₂CHClCH₂CHCl ∼∼ ⟶ HCl⁺ ∼∼ CH₂CHClĊHCHClCH₂CHCl ∼∼

⟶ C··⁺ ∼∼ CH₂CHClCH=CHCH₂CHCl ∼∼

⟶ Repeat

Chain termination

$$\left.\begin{array}{c} Cl* + Cl* \\ R* + Cl* \\ R* + Cl* \end{array}\right\} \longrightarrow Product$$

Fig.7.14 Radical mechanism for the dehydrochlorination of PVC

∼∼CH₂CHClCH₂CHClCH₂CHCl∼∼ ⟶ ∼∼CH₂ĊH⁺—C̄HCHClCH₂CHCl∼∼ ⟶
           Cl⁻---H⁺

⟶ ∼∼CH₂CH=CHCHClCH₂CHCl∼∼ + HCl ⟶ ∼∼CH₂CH=CH—ĊH⁺—C̄HCHCl∼∼
                                                          Cl⁻--H⁺

⟶ ∼∼CH₂CH=CHCH=CHCHCl∼∼ + HCl ⟶ Futher release HCl

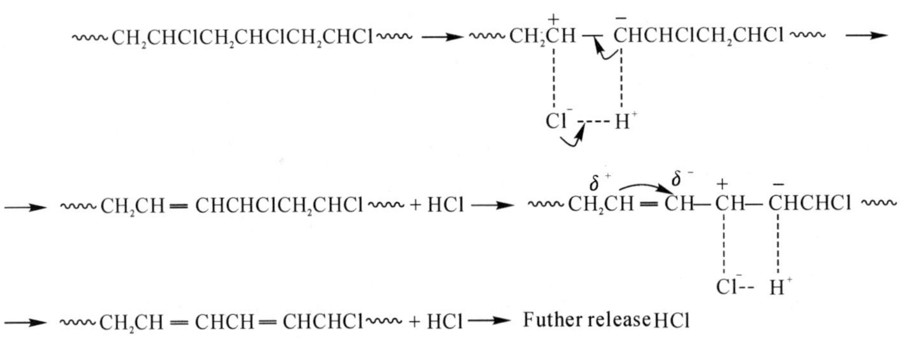

Fig.7.15 Ion-molecular mechanism for the dehydrochlorination of PVC

3) Molecular mechanism. According to this mechanism, the dehydrochlorination reaction undergoes through a four member transition state, and the dehydrochlorination reaction catalyzed by HCl goes through a six element transition state (see Fig. 7.16). In fact, the dehydrochlorination reaction of PVC may be carried out by several mechanisms at the same time.

(3) Crosslinking reaction. The conjugated polyene structure is formed after the degradation and dehydrochlorination of polyvinyl chloride, and this structure may undergo intermolecular Diels-Alder reaction (see Fig. 7.17). The intramolecular cyclization reaction of polyene structure will lead to the formation of benzene and other aromatic structures, which is also a possible route to form carbonized residues at higher temperatures.

(4) Oxidative chain breaking. In the presence of oxygen, the free radical degradation of PVC will inevitably occur oxidation reaction. Its oxidative chain breaking process is similar

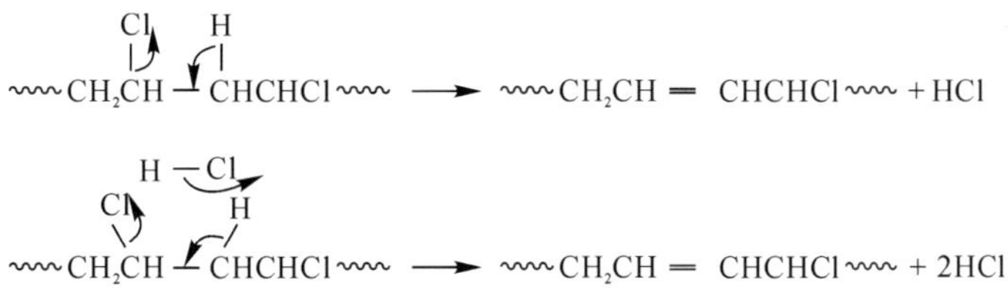

Chapter 7 Degradation and stabilization of different class of commodity polymers

Fig.7.16 Molecular mechanism for the dehydrochlorination of PVC

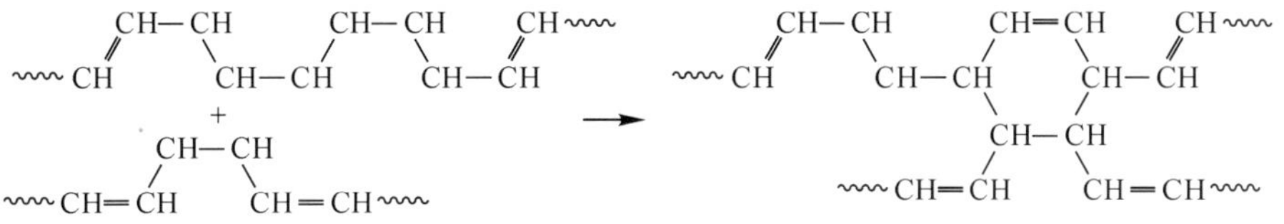

Fig.7.17 Diels-Alder reaction of conjugated polyene compounds during the degradation of PVC

to that of polyolefin. Macromolecular free radicals interact with oxygen to generate peroxyl free radicals. The latter abstracts hydrogen atoms and can be converted into hydroperoxides. Hydroperoxides decompose to generate macromolecular alkoxy free radicals, and finally lead to macromolecular chain breaking (see Fig. 7. 18).

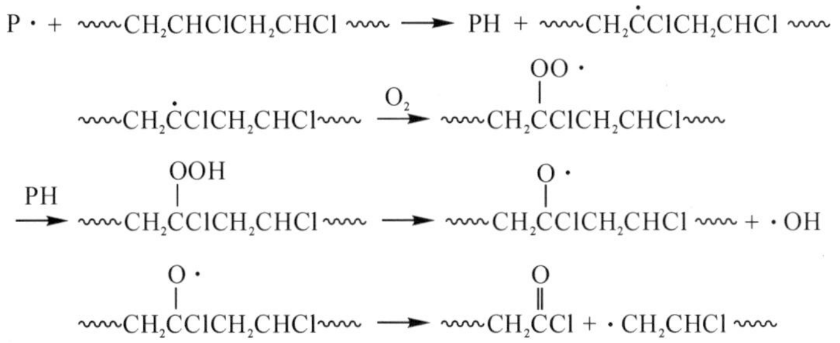

Fig.7.18 Oxidation induced chain breaking of PVC

Alkoxy radicals can also undergo the following reactions to form carbonyl group with chlorine atom at β-position. This combination activated the dehydrochlorination of polyvinyl chloride and generated a carbonyl allyl structure (see Fig. 7. 19). This structure and allyl structure play an important role in the dehydrochlorination process of PVC.

2. Photo degradation of polyvinyl chloride

Polyvinyl chloride is degraded and crosslinked during UV irradiation, and conjugated polyenes and hydrogen chloride are also generated. Dehydrochlorination reaction changed the absorption spectrum of polyvinyl chloride, and the polyene structure formed made polyvinyl chloride colored. It is generally believed that the mechanism of photodegradation of polyvinyl chloride is free radical mechanism. The first step is to randomly break the chain to generate free radicals, as shown in reaction formula (a) in Fig. 7. 20. The second step is to generate

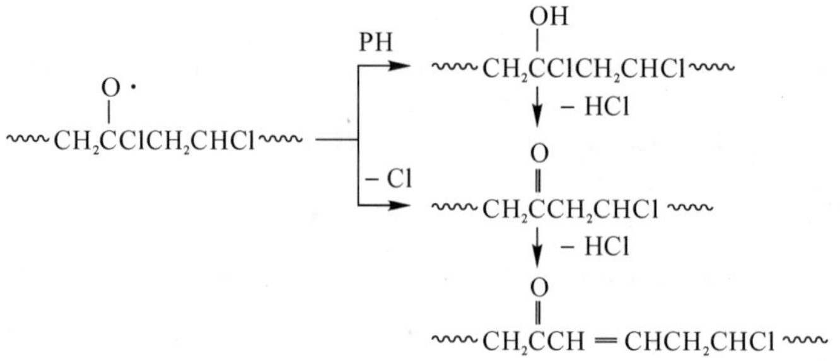

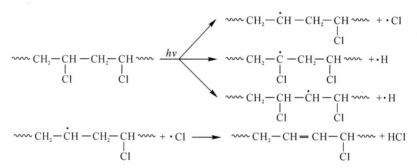

Fig.7.19 The formation of carbonyl allyl group during the degradation of PVC

an isolated unsaturated bond on the main chain, which is likely to be completed by chlorine radicals attacking macromolecular free radicals, as shown in reaction formula (b) in Fig. 7. 20.

Fig.7.20 Release of HCl during the photo degradation of PVC

It is generally accepted that dehydrochlorination reaction is a "zippering" reaction. To make the color of the polymer yellow, at least seven conjugated double bonds are required. Sequential removal of HCl from macromolecules means increasing the conjugation energy of the remaining chains, which reduces the activation energy required for the next step of dehydrochlorination, so makes it easy to form polyene chains.

Many people found that HCl can accelerate the photodegradation of PVC. However, only when there are conjugated polyenes on the PVC chain can there be significant HCl acceleration effect. Therefore, in addition to the free radical mechanism, the dehydrochlorination reaction during the photodegradation of polyvinyl chloride may also have a molecular mechanism and an ionic mechanism. Gibb reported that when PVC was irradiated with ultraviolet light, the degradation reaction was only carried out on the thin surface of the sample. The experimental results show that the degradation reaction is confined to the surface layer with a thickness of about 0. 2 mm. The amount of HCl escaping is proportional to the surface area of the sample, but independent of the film thickness. This result also indicates that the surface reaction is carried out. In the first 1 h of the reaction, the reaction rate will depend on the light intensity and temperature, but after that, the reaction rate will be independent of these two parameters. When irradiated in air, the gel content is lower than that in nitrogen, which may mean that there is competition between oxidation reaction and crosslinking reaction. In the presence of air, dehydrochlorination and oxidation reaction occur simultaneously in the photo induced degradation process.

Polyvinyl chloride can be degraded under ultraviolet irradiation with wavelength larger than 280 nm, which indicates that there may be carbonyl groups on the polymer chain, and there may also be some impurities in the polymer, such as trace solvents remaining in the polymerization process. Kamal reported that the UV resistance of PVC films containing trace solvents was significantly reduced. When the residual solvents (such as tetrachlorofuran and dichloromethane) are increased during coating, the degradation rate and discoloration rate of the polymer increase under 300 nm irradiation.

Bellenger found that orientation has an important effect on the photo-oxidation of PVC. After stretching, it will generate much more carbonyls under the irradiation of sunlight ultraviolet (wavelength greater than 300 nm) than the un-stretched samples (see Fig. 2.21). The photosensitivity of the samples increases with the increase of orientation (characterized by infrared dichroism). This is explained as a conformation formed by PVC during stretching, which is favorable for free radicals to capture hydrogen atoms from the molecular chain.

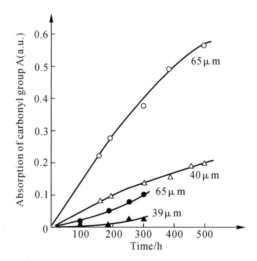

Fig. 2.21 The relationship of carbonyl content and irradiation time during the photo degradation of PP film: ○, △ for stretched sample; ●, ▲ for un-stretched sample

Recently, Carlsson and coworkers have studied the degradation of rigid PVC sheets under UV and sunlight exposure, and measured the change of the rate of volatiles production with the time of UV irradiation. The results showed that the rate of volatiles production decreased rapidly with the extension of UV irradiation time. The authors believe that this may be due to the cross-linking of the irradiated sample surface caused by degradation, which makes the sample surface lack chlorine and rich in titanium dioxide pigment (Rutile type). The surface of the sample irradiated by sunlight is not crosslinked, and the released volatile products are similar to those of the sample irradiated in the laboratory. Horikoshi and others studied the photodegradation of PVC in aqueous suspensions. Two kinds of PVC samples were tested, one is the alloy of PVC/TiO_2 mixed uniformly, and the other is the suspension of PVC particles or films and TiO_2. The results show that the alloy sample is easier to degrade than the suspension sample. Remillard and others studied the degradation of PVC during outdoor aging and aging in aging box by Raman spectroscopy and fluorescence spectroscopy. Because the resonance behavior of polyenes is closely related to the length of their conjugated sequences, Raman spectra and fluorescence spectra are very sensitive to the identification of conjugated sequences with a length of about 10 – 20. The analysis of the samples aged at 100 – 120 °C for 500 h after 35 months outdoor weathering and aging in the aging box shows that the integration of the fluorescence spectrum of the outdoor weathering samples has good coherence with the length of the conjugated polyene sequence measured by the Raman spectrum, indicating that the increase in the number of short and long conjugated polyene sequences is similar. The samples in the aging chamber first form short conjugated polyene sequences, which develop into long sequences with the degradation.

7.2.2 Degradation of other chlorine-containing polymers

1. Polyvinylidene chloride

The thermal stability of polyvinylidene chloride (PVDC) is slightly worse than that of PVC and it degrades to release HCl at 120 – 220 °C. When the temperature is higher than 220 °C, other degradation products are also produced. Each vinylidene chloride unit produces only one molecule of HCl. PVDC changes color quickly, turning yellow when 1 percent volatiles are produced, and turning dark brown when 10% volatiles are produced. It becomes completely insoluble when 1% volatiles are produced by degradation. As for the mechanism of dehydrochlorination of PVDC, the free radical mechanism of "zippering" reaction initiated at the chain end and carried out along the molecular chain was proposed.

Chain breaking and cross-linking of vinyl chloride vinylidene chloride copolymer occur during photolysis, and hydrogen chloride is generated from vinylidene chloride unit.

2. Chlorinated rubber

The chlorine content of commercial chlorinated rubber is 65% to 68%. Chlorine is bound to the molecular chain of polymer through addition or substitution reaction. The structure of chlorinated rubber is very complex, and the molecular chain also includes some cyclic units. The thermal degradation of chlorinated rubber was studied by TG and TVA. When the temperature is increased at the rate of 10 °C/min, the chlorinated rubber is still stable until 200 °C, and the weight loss is carried out in two stages. The maximum rate of weight loss in the first stage is about 300 °C, and the weight loss in this stage is about 67%. The weight loss in the second stage occurs between 400 °C and 500 °C, and the weight loss is only 3%. In the first degradation stage, almost 95% of the volatile degradation products are

HCl, with a small amount of CO and CO₂ and trace amounts of CH_4, C_2H_4. The volatile products of the second stage degradation are CH_4, C_2H_4, HCl and H_2. When the dehydrochlorination of chlorinated rubber reaches 1%, it changes color, indicating that the conjugated double bond structure is formed on the molecular chain as in the degradation of polyvinyl chloride. At this degradation level, the polymer becomes insoluble.

3. Polychloroprene

Unless a stabilizer is added, polychloroprene will absorb oxygen and become unstable in the air. It absorbs oxygen and releases hydrogen chloride. The rate of releasing hydrogen chloride is very close to its oxygen absorption rate. The effect of heat on polychloroprene must be studied in nitrogen or in vacuum. Thermogravimetric analysis shows (see Fig. 7.22) that, like PVC, the weight loss of polychloroprene can be divided into two stages, but the thermal decomposition temperature is much higher than that of PVC, and the carbonization residue left at 500 ℃ is much more than that of PVC (up to 20% of the sample mass). Comparing the thermogravimetric data with the amount of HCl released (see Fig. 7.22), it is obvious that there are other degradation products besides HCl. The weight loss in the first stage is about 45%, but only about 35% is HCl. This amount is less than 90% of HCl in polychloroprene calculated according to the chlorine content.

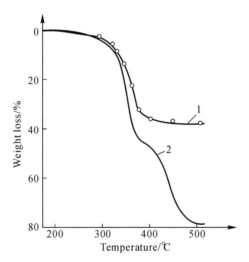

Fig.7.22 The relationship between amount of HCl released during the thermal degradation of polychloroprene and TGA result: 1 for HCl release; 2 for TGA

It is found that the degradation gas product contained a small amount of ethylene and trace amounts of monomers. Above 400 ℃, methane, a small amount of hydrogen and propylene are produced. In addition to gas products, there are complex liquid products, including two isomers of dichloro-4-vinyl cyclohexene.

The UV absorption spectra of polychloroprene samples with different dehydrochlorination degrees were studied, the main absorption is related to the triene structure. This may be due to the random dehydrochlorination process of non free radical mechanism (see Fig. 7. 23). In the polychloroprene/PMMA mixture, polychloroprene did not accelerate the degradation of PMMA, which also proved that dehydrochlorination was a non radical reaction.

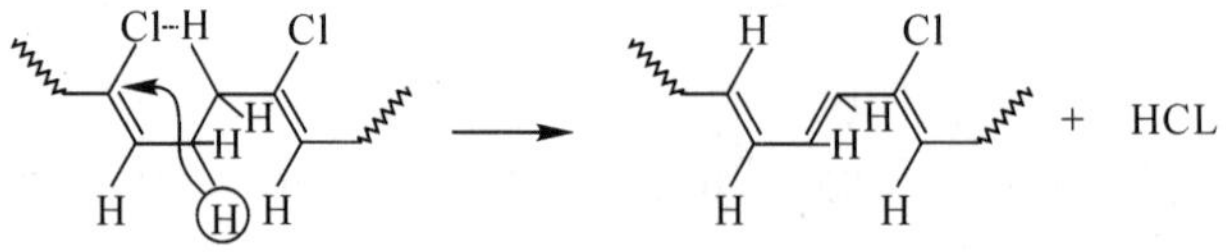

Fig.7.23　Mechanism for the release of HCl during the degradation of polychloroprene

4. Other chlorinated polymers and copolymers

Chlorinated polyvinyl chloride crosslinks quickly and degrades under UV irradiation. When the content of vinyl fluoride in vinyl chloride-vinyl fluoride copolymer film increases, the degradation resistance increases.

7.3　Degradation and stabilization of polystyrene and styrene copolymers

7.3.1　Degradation of polystyrene

1. Thermal degradation of polystyrene

The degradation behavior of polystyrene is closely related to whether the temperature exceeds 300 ℃. Between 200 ℃ and 300 ℃, the relative molecular weight of the polymer decreases, but there is no volatile product. The relationship between the decrease of relative molecular weight and time during the degradation of polystyrene indicates that random chain breaking occurs. The reasons are as follows.

Assuming that the average number of breaks of each molecule at time t is S, P_0 and P_t are the initial length of the polymer molecular chain and the length of the time, respectively, then

$$P_t = P_0/(S+1)$$

That

$$S = (P_0/P_t) - 1$$

Chain breaking fraction α defined as the number of breaks per molecule divided by the initial length of the molecule,

$$\alpha = S/P = (1/P_t) - (1/P_0)$$

For the random broken chain, the breaking of each bond is similar, so there are

$$\alpha = k \times t$$

Therefore, a plot of time should be a straight line through the origin. The degradation curve of polystyrene obtained by anionic polymerization is shown in Fig. 7.24. However, the degradation curve of polystyrene obtained by radical polymerization is a straight line that does not pass through the origin, which indicates that there is a subordinate relationship:

$$\alpha = \beta + kt$$

β indicates the initial defect existing in the sample, which may be peroxide.

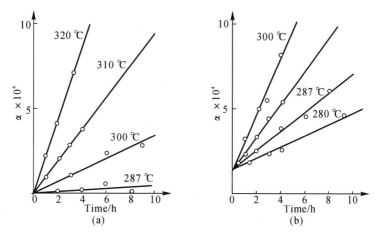

Fig.7.24 Relationship between the chain breaking fraction α and time during the thermal degradation of polystyrene: (a) for PS prepared from anionic polymerization, $M_n = 22.9$ kg/mol; (b) for PS prepared from free radical polymerization, $M_n = 1,490$ kg/mol

The mechanism of thermal degradation process below 300 ℃ may be: after the initial molecular chain is split, the generated free radicals A and B are terminated by disproportionation due to the cage effect (see Fig. 7.25). The depolymerization does not occur in this temperature range, which also indicates that the C and D formed are stable up to 300 ℃.

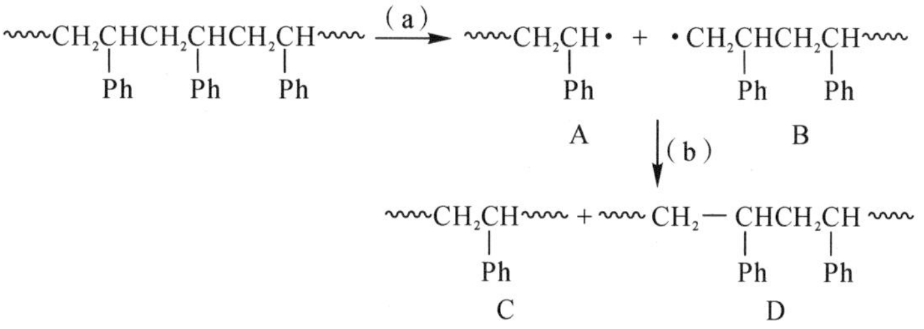

Fig.7.25 Random chain breaking and disproportionation for PS

When the temperature is higher than 300 ℃, the degradation products of polystyrene are various low-molecular compounds. The main components are monomers (monomer yield is 40% to 60% by mass), dimers, trimers, and a small amount of toluene (2%) and α-methylstyrene (0.5%). There are CRF (cold ring fraction) components (dimer and larger fragments) in PS degradation products, which clearly indicates that there are end group benzene C and unsaturated end group D in the degradation process.

After the tertiary hydrogen atom in polystyrene is replaced with deuterium, the yield of monomer during degradation is greatly improved. Connect to the fact that 100% monomers are obtained during the degradation of poly α-methyl styrene clearly illustrates the importance of the chain transfer reaction of tertiary hydrogen atoms to the formation of non-volatile degradation products.

When polystyrene is degraded at a temperature higher than 300 ℃, its initiation process may be random homocracking of the molecular chain, reaction (a) in Fig. 7.25, or homocracking near the C-type or D-type chain end (see Fig. 7.26).

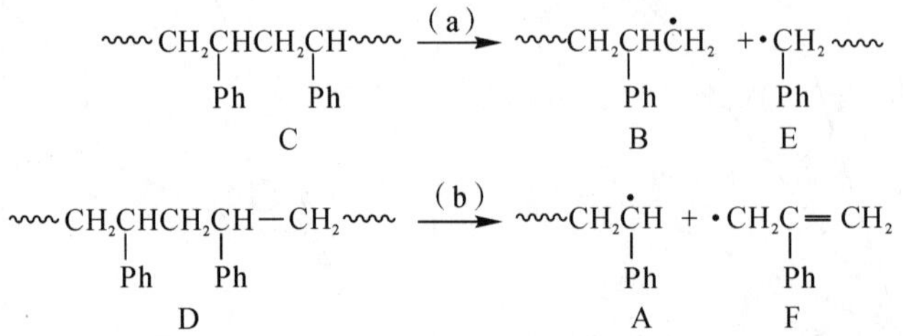

$$\underset{\underset{C}{Ph\quad Ph}}{\sim\sim\sim CH_2CHCH_2CH\sim\sim\sim} \xrightarrow{(a)} \underset{\underset{B}{Ph}}{\sim\sim\sim CH_2CHCH_2} + \underset{\underset{E}{Ph}}{\cdot CH_2\sim\sim\sim}$$

$$\underset{\underset{D}{Ph\quad Ph}}{\sim\sim\sim CH_2CHCH_2CH-CH_2\sim\sim\sim} \xrightarrow{(b)} \underset{\underset{A}{Ph}}{\sim\sim\sim CH_2\dot{C}H} + \underset{\underset{F}{Ph}}{\cdot CH_2C=CH_2}$$

Fig.7.26 The chain breaking reaction for the chain end during the degradation of PS

Only the formation of free radicals, as shown in Fig. 7.26 A and B, will lead to the direct generation of volatile degradation products (see Fig. 7.27). Free radical A is depolymerized in reaction (a) to form monomers, while intramolecular chain transfer reactions [reactions (b) and (c)] lead to the formation of dimers, etc. Fragments such as trimers are generated in a similar manner, that is, by the transfer reaction of the next tertiary hydrogen atom in the molecular chain.

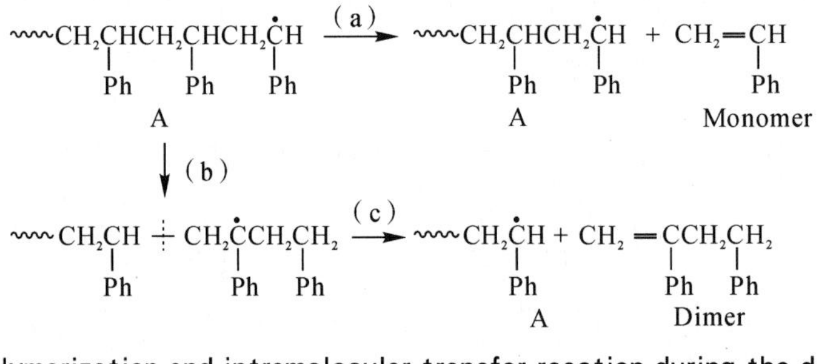

$$\underset{\underset{A}{Ph\quad Ph\quad Ph}}{\sim\sim\sim CH_2CHCH_2CHCH_2\dot{C}H} \xrightarrow{(a)} \underset{\underset{A}{Ph\quad Ph}}{\sim\sim\sim CH_2CHCH_2\dot{C}H} + \underset{\underset{Monomer}{Ph}}{CH_2=CH}$$

$$\downarrow (b)$$

$$\underset{\underset{}{Ph\quad Ph\quad Ph}}{\sim\sim\sim CH_2CH \dotplus CH_2\dot{C}CH_2CH_2} \xrightarrow{(c)} \underset{\underset{A}{Ph}}{\sim\sim\sim CH_2\dot{C}H} + \underset{\underset{Dimer}{Ph\quad Ph}}{CH_2=CCH_2CH_2}$$

Fig.7.27 Depolymerization and intramolecular transfer reaction during the degradation of PS

Therefore, all volatile degradation products are produced by the further reaction of free radical A. The ratio between monomers and other volatile products depends on the competition between reactions (a) and (b) in Fig. 7.27. The kinetic chain length of radical a is about 50 units, which is much smaller than that of PMMA and PAMS.

The continuous decrease of molecular weight during PS degradation is due to the random breaking of molecules or the chain transfer between molecules, which is particularly important for the reduction of relative molecular mass for chain transfer. Each free radical A, B, E and F can be used as R· in reaction (a) of Fig. 7.26 to capture hydrogen atoms. The generated tertiary carbon atom macromolecular radical g generates β-fracture, and then generates a radical A and an unsaturated end group D. Reaction (a) in Fig. 7.28 may be the only important reaction of free radical B to generate a methyl terminal group. Radical a

Chapter 7 Degradation and stabilization of different class of commodity polymers

reacts in this way to form a phenyl terminal group C. A small amount of toluene and α-methylstyrene is produced by the free radicals E and F through the reactions (a) and (b) in Fig. 7.27, and then through the reaction (a) in Fig. 7.28 to seize a hydrogen atom.

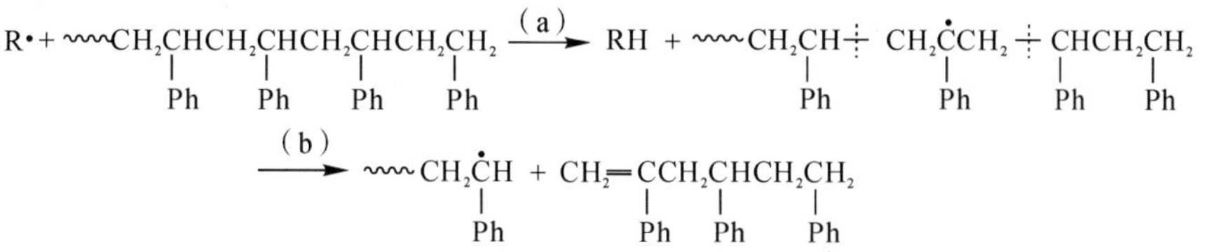

$$R\cdot + \sim\sim CH_2CHCH_2CHCH_2CHCH_2CH_2 \xrightarrow{(a)} RH + \sim\sim CH_2CH \dotplus CH_2\dot{C}CH_2 \dotplus CHCH_2CH_2$$
(with Ph groups on each substituted carbon)

$$\xrightarrow{(b)} \sim\sim CH_2\dot{C}H + CH_2=CCH_2CHCH_2CH_2$$
(with Ph groups on each substituted carbon)

Fig. 7.28 Intermolecular transfer reaction and chain breaking during the degradation of PS

Many people have thought that unsaturated end group D triggers depolymerization through reaction (b) in Fig. 7.27, but the radical a formed by random chain breaking, especially the subsequent intermolecular chain transfer reaction, is the basic route to produce volatile degradation products. The relative yields of toluene and indeed very small amounts of α-methylstyrene also support this view. It is also shown that the phenyl terminal group C may be more unstable than the unsaturated terminal group D. In the early stage of PS degradation with only phenyl end group, more toluene was produced, further indicating that the degradation reaction was initiated from this end group.

McNeill and others analyzed the degradation products above 300 ℃ in detail. Table 7.2 shows the degradation products of an anionic polymerized PS at three temperatures (in order of importance of the products). Styrene is the main product at all three temperatures. Toluene was found at all three temperatures, indicating random fracture at all three temperatures. The α-methylstyrene is derived from the vinylene end group structure, which is generated by the termination of disproportionation after random fracture, or by the cleavage of a large fraction of free radicals. During the degradation process, macromolecular primary radicals are transferred into secondary radicals through intramolecular chain and released simultaneously α-methylstyrene or CRF containing 5 - 7 carbon atoms in the main chain. 1-methylindene existed in the degradation at 350 ℃ but not in the degradation product at 420 ℃, the 3-methylindene only exists in the degradation products at 350 ℃, which is attributed to the degradation of the head-to-head structure of PS initiated by sodium naphthalene. There was 3-phenylpropene at 300 ℃ and more at 350 ℃, but there was no such product at 420 ℃. There are also some products at 350 ℃, which neither appear at 300 ℃ nor at 420 ℃. The ratio of volatile products to CRF is not constant. Through analysis, the author concludes that at relatively low temperature, the intramolecular chain transfer (generation of CRF) is limited due to the high melt viscosity, so the intermolecular chain transfer reaction mainly occurs, which leads to the reduction of the relative molecular weight of the polymer. After 25% weight loss, intramolecular chain transfer becomes the main chain transfer reaction, so more CRF products are produced.

Table 7.2 Volatile products produced at different temperatures during the thermal degradation of PS prepared from anionic polymerization

300 ℃	350 ℃	420 ℃	300 ℃	350 ℃	420 ℃
Styrene	Styrene	Styrene	1-methylindene	Benzene	
Toluene	Toluene	Toluene	3-phenylpropene	Dimethylindene	
Benzene	α-methyl styrene	Ethylbenzene		trans-2-methylstytene	
Naphthalene	1-methylindene	α-methyl styrene		3-methyllindene	
α-methyl styrene	3-phenyl propene			4-phenylbutene-1	

Nyden simulated the thermal degradation of polystyrene by computer through molecular dynamics model and found that the main reactions include β-breaking, random chain breaking and C—C bond breaking between tertiary and secondary carbon atoms. The latter leads to the formation of benzyl radicals and toluene, random chain breaking leads to the formation of styrene oligomers, and β-fracture leads to the formation of styrene monomer. The degradation products such as benzene, ethylene and acetylene were also observed in the simulation results. The head to head structure only affected the degradation at relatively low simulation temperature. At this time, the main reaction was the breaking of the head to head bond.

2. Photo degradation of polystyrene

The UV absorption of polystyrene is caused by the transition from the ground state to the excited state of the benzene ring, and the rest of the polymer molecule does not absorb light with a wavelength greater than 200 nm. The non degraded polystyrene has absorption up to 280 nm, and the absorption spectrum of the degraded polystyrene is extended to 340 nm.

The first step of photodegradation is that the benzene ring absorbs light to generate the benzene ring in the excited singlet state, and then it is transformed into the triplet state through the inter system crossing. The second step is the reaction of the triplet state of the benzene ring, including: ① the energy of the triplet excited benzene ring can be used to dissociate the C_6H_5—C bond. ② through intramolecular energy transfer, the triplet excitation energy can be transferred to the C—H bond or the C—C bond, breaking the two bonds.

The most important step of photolysis of polystyrene in vacuum is the breaking of C—H bonds, especially C—H bonds on tertiary carbon atoms. Hydrogen radicals are easy to

Chapter 7 Degradation and stabilization of different class of commodity polymers

move. They can diffuse out of the polymer body and combine with each other to form hydrogen molecules. Because benzene radicals are limited by their movement ability, they cannot diffuse out and can only react around themselves. The most likely reaction is that benzene radicals take hydrogen atoms from polymer molecules.

Polystyrene was photo decomposed with 254 nm light in vacuum, and the only gas product was hydrogen. In the solid state, the movement of polymer macromolecules is limited, but free radicals can move along the polymer chain until they are captured by other free radicals or impurities. Crosslinking may occur when two macromolecular radicals are close to each other. In addition, macromolecules can also undergo disproportionation fracture (β-fracture). When terminal free radicals are generated, zippering reaction may occur to generate monomers. Polystyrene samples showed an increase in optical density during UV irradiation and slightly yellowish. Grassie believes that the discoloration of polystyrene is due to the formation of conjugated double bonds on the main chain of the sub polymer. However, Rabek believes that the yellowing of polystyrene in vacuum is due to the isomerization of benzene molecules in photodegradable polymers or benzene rings in polystyrene into benzofulvene and benzylidene. The yellow color of benzofulvene is due to its conjugated double bond system. When polystyrene is irradiated in nitrogen atmosphere, the yellowing rate will increase, but the mechanism of this effect is not clear.

The mechanical properties of polystyrene film in air (or oxygen) change rapidly under ultraviolet irradiation, and the film becomes brittle and seriously yellow. The main reactions observed were chain breaking, crosslinking and oxidative degradation. The rate of photooxidation is independent of the relative molecular weight of the polymer, but is proportional to the oxygen partial pressure. The rate of photooxidation is also related to the wavelength. The rate of oxygen uptake is much slower when irradiated with light at 365 nm than when irradiated with light at 254 nm, and the reaction shows a typical induction period.

The photo oxidation mechanism of polystyrene can be divided into the following categories:

(1) Initiation reaction. In the ultraviolet region with wavelength less than 280 nm, the reaction may start with the breaking of C—H bond or C—C bond to generate macromolecular radicals. Rabek and others proposed the mechanism of photooxidation initiation stage and the mechanism of the reaction between basic singlet oxygen and polystyrene molecule. The mechanism of initiation under the action of light with wavelength greater than 280 nm is still unclear.

(2) Kinetic chain growth reaction. The following reactions can be carried out, as shown in Fig. 7.29. The polymer radicals generated in the initiation react with oxygen molecules to generate polymer peroxyl radicals [reaction (a)], and the latter reacts with surrounding polystyrene molecules to capture hydrogen atoms [reaction (b)]. The polymer hydroperoxide decomposes under light irradiation [reaction (c)], the polymer alkoxy radicals produced by the decomposition of hydroperoxide can be β-decomposition by fracture

[reaction (d)]. In addition, the polymer free radicals can also be broken through disproportionation [reaction (e)].

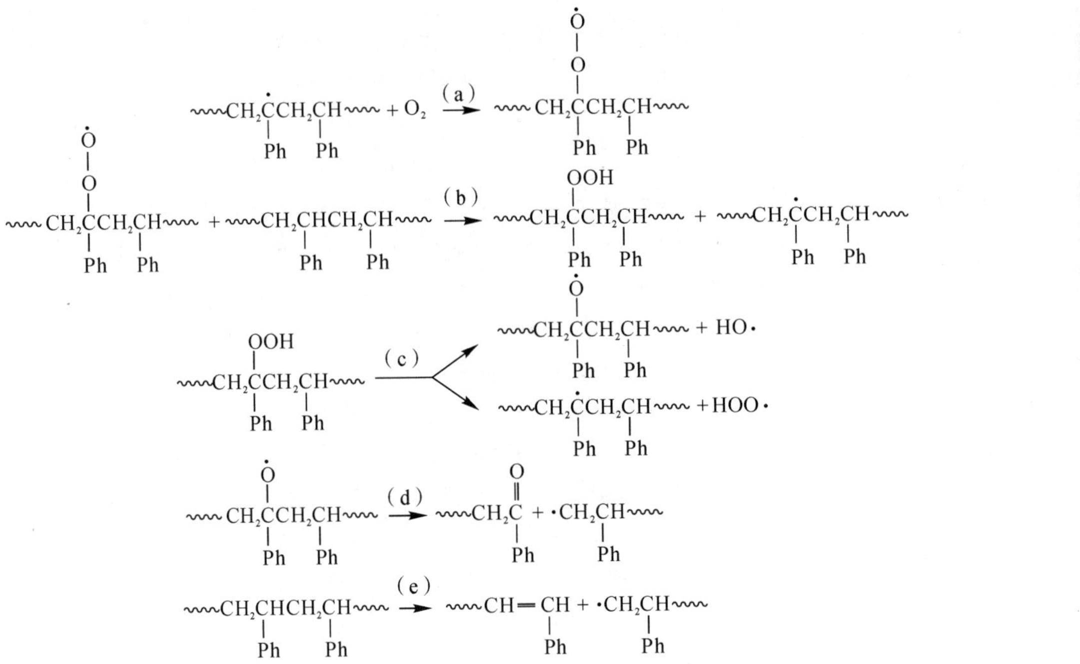

Fig.7.29 Kinetic chain growth reaction during the photo-oxidation degradation of PS

(3) Chain termination reaction. The termination of the kinetic chain is due to the binding of free radicals into inactive products. During the photo-oxidation of polystyrene, the products volatilized from the polymer are hydrogen, carbon monoxide, carbon dioxide, water, methanol, benzene, styrene, styrene and benzophenone.

Two characteristic infrared absorption bands were formed during the photooxidation of polystyrene. The band of hydroxyl group is $3,600 - 3,400$ cm^{-1} and the band of carbonyl group $1,800 - 1,700$ cm^{-1}. Rabek reported that the maximum absorption of the carbonyl absorption band was $1,740$ cm^{-1}, which could not be explained by the formation of acetophenone, but it was thought that the ring opening reaction of the benzene ring in the polystyrene molecule produced the conjugated dialdehyde group. However, this reaction occurs only in the presence of oxygen in the process of illumination. Such explanation is based on:

1) The infrared spectrum of the final product of photo-oxidation of liquid benzene is very similar to that of polystyrene film after photo-oxidation (see Fig. 7.30).

2) The polystyrene film formed an absorption band at 274 nm under UV irradiation, while the *trans, trans*-2, 4-hexadiene-1, 6-dialdehyde obtained by benzene ring opening reaction also had an absorption peak in the same spectral region.

When the polystyrene dissolved in the deoxygenated benzene was irradiated with ultraviolet light, it was found that the viscosity of the solution did not change, and no chain

Chapter 7　Degradation and stabilization of different class of commodity polymers

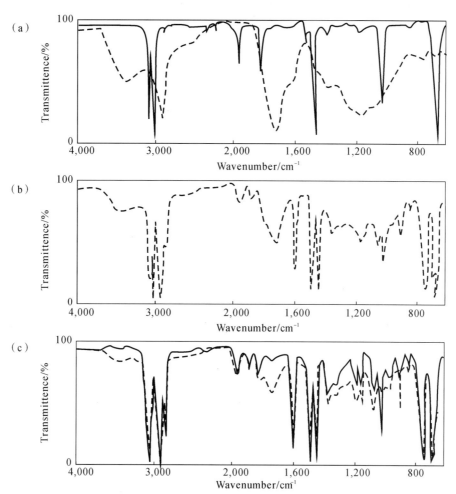

Fig. 7.30　FT-IR spectra for the photo oxidation degradation of benzene and PS: (a) FT-IR spectra of benzene in C_{14} (solid line) and liquid benzene after 10 h of UV irradiation in KBr pellet; (b) FT-IR spectrum of the product after 5 h UV irradiation of liquid benzene (non-irradiated PS as container); (c) FT-IR spectra of degraded PS film (solid line) and PS film after 5 h UV irradiation (dash line)

breaking and cross-linking were observed. Rabek pointed out that the benzene molecules in the deoxygenated benzene solution act as an internal filter, which absorbs the incident ultraviolet light. Then, the excited singlet and triplet benzene molecules are quenched by other molecules in the ground state. Once a polystyrene molecule absorbs ultraviolet light, it is also rapidly deactivated by the surrounding benzene molecules. After the polystyrene solution is saturated with air or oxygen and then irradiated with ultraviolet light, the viscosity of the solution decreases rapidly (see Fig. 7.31). The film made of this polystyrene solution is yellow and brittle. These films can be completely dissolved in chloroform, and even no trace of gel can be seen, indicating that there is no cross-linking in the solution.

In benzene solution, polystyrene can be degraded by light with wavelength greater than 280 nm (pure polystyrene does not absorb light in this wavelength band). Rabek and others

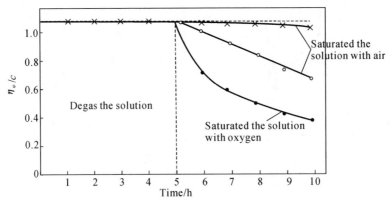

Fig.7.31 Viscosity changes during the UV degradation of PS under different conditions

hypothesized that a charge transfer complex (CTC) was generated between polystyrene and molecular oxygen. The formation of this complex can explain the absorption of polystyrene in 280 – 340 nm and longer wavelength range. It is likely that this photo excited complex can be decomposed to generate singlet oxygen, and then the latter can participate in the photoinitiated reaction of the oxidation process (see Fig. 7.32).

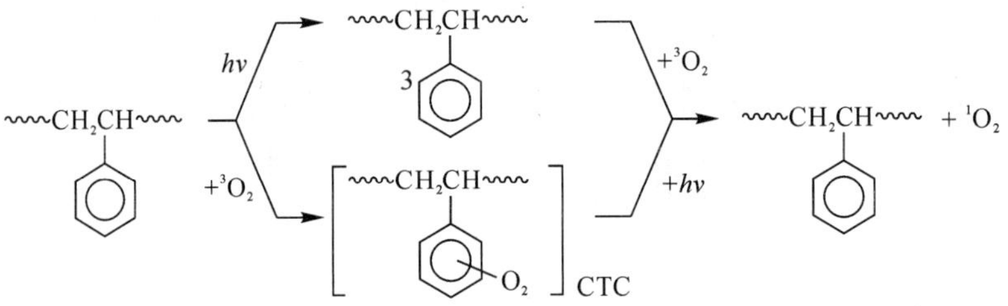

Fig.7.32 CTC formed between PS and O_2

7.3.2 Degradation of poly α-methylstyrene

1. Thermal degradation of poly α-methylstyrene

The degradation behavior of poly α-methylstyrene (PAMS) is very different from that of PS, and its thermal stability is much worse than that of PS. The volatile products began to be released at about 250 ℃, and the reaction was rapid at 300 ℃, and the yield of monomers was almost 100%. The relative molecular weight of PAMS decreased much less than that of PS. The thermal degradation study in solution showed that the initiation rate of the degradation reaction was proportional to the relative molecular weight of the polymer.

The mechanism of PAMS degradation can be shown in Fig. 7.33. After initiation of random chain break [reaction (a)], depolymerization reaction [reaction (b)] occurs. It was once thought that the macromolecular radicals A and B generated by the random end chains all undergo depolymerization to form monomers. In fact, the free radical a must be depolymerized into a monomer in this way, because it is the growth chain free radical during the polymerization of α-methylstyrene, and the maximum polymerization temperature of α-methylstyrene is very low. However, radical B, like the radicals generated during PS degradation, tends to form stable molecules by capturing hydrogen atoms.

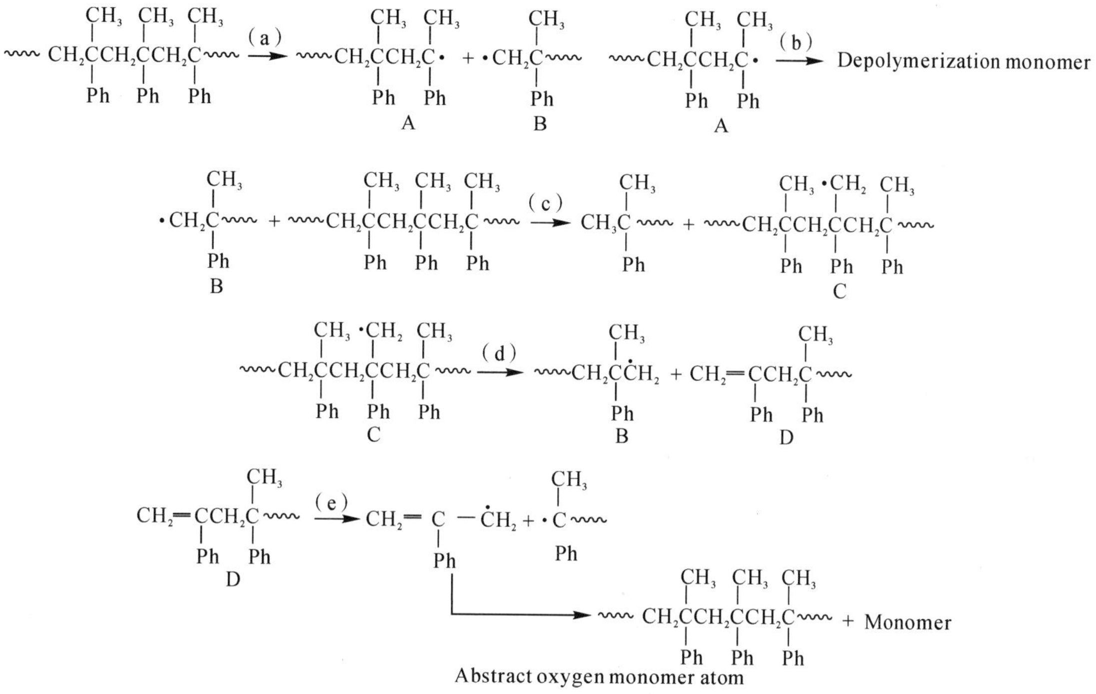

Fig.7.33 Thermal degradation of PAMS

The experimental data of PAMS degradation were compared with the computer simulation data. The computer simulation is based on the fact that the length of the free radicals in the depolymerization reaction is longer than the kinetic chain. Computer simulation is carried out in two cases: ① both radicals generated by chain initiation depolymerize to form monomers. ② only one radical depolymerizes, while the other radical becomes a stable molecule through chain transfer. The computer simulation also simulated different kinetic chain lengths, and the results are shown in Fig. 7.34. The figure clearly shows that the simulation curve with only one free radical depolymerizing reaction is better consistent with the experimental data. When the kinetic chain length is 2,000, the degradation of polymers with a wide range of relative molecular mass can be well simulated.

2. Photo degradation of poly α-methylstyrene

The photodegradation reaction of poly α-methylstyrene is very different from that of polystyrene. During photolysis in vacuum, the first radical generated by polystyrene is easily combined with the second radical generated, resulting in crosslinking and stabilization. Or disproportionation reaction with the second radical to form a double bond without main chain

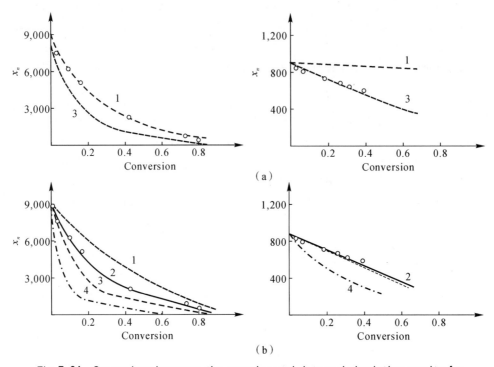

Fig.7.34 Comparison between the experimental data and simulation results for the relationship between the DP and conversion during the degradation of PAMS with higher and lower molecular weight: (a) degradation induced by the two radicals generated by the chain breaking; (b) degradation induced by one of the radicals generated by chain breaking. 1 for kinetic chain length of 5,000, 2 for 2,000, 3 for 425, 4 for 324

breaking. In the case of poly α-methylstyrene, the relative molecular weight decreases rapidly, and monomers are formed accordingly. At the initial stage of degradation, the sharp decrease of the relative molecular weight usually indicates the random fracture of the polymer chain. Poly α-methylstyrene is not easy to crosslink in solid state or solution. The photodegradation reaction in vacuum can be shown in Fig. 7.35. Reaction (a) is the initiation reaction and (b) is the depolymerization reaction. The diffusion of monomer in the polymer film and the volatilization from the surface are very slow. It is likely that some macromolecular radicals have carried out hydrogen transfer reaction with monomer molecules [reaction (c)].

Ploy α-methylstyrene produces volatile products after long-term irradiation at 115 ℃, and the main volatile products are hydrogen and carbon dioxide. The hydrogen atoms formed after the C—H bond is broken combine with each other to form hydrogen molecules. The generation of carbon dioxide during illumination indicates the existence of oxidizing groups in the polymer.

Yellowing of samples was observed during irradiation of poly α-methylstyrene in solid state or benzene solution with ultraviolet light in the presence of air or oxygen. The infrared spectrum of poly α-methylstyrene irradiated by ultraviolet light shows that the characteristic absorption bands produced when the poly α-methylstyrene is oxidized and opened with benzene molecules are formed in $1,740$ cm^{-1} and $3,400$ cm^{-1} region, which indicates that the yellowing of poly α-methylstyrene samples is caused by the photo oxidative ring opening of benzene rings on polymer molecules.

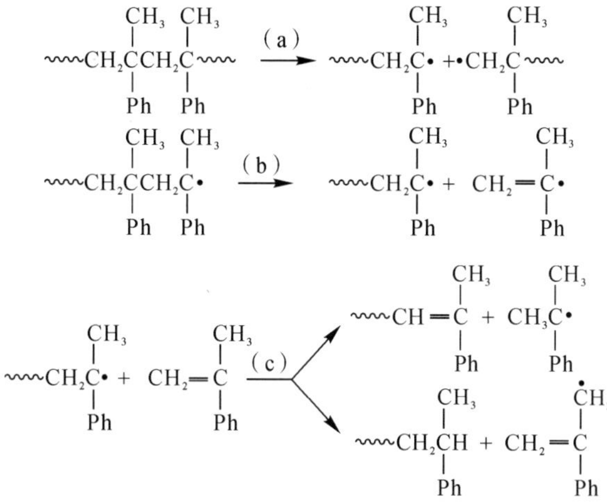

Fig.7.35 Photo degradation of PAMS in vacuum

7.3.3 Degradation of styrene copolymer

The thermal stability of styrene/MMA (S/MMA) copolymer with styrene as the main component is between the homopolymers of the two monomers. However, the type and content of degradation products are greatly affected by MMA. For example, for the copolymer with S/MMA of 4:1, the yield of nonvolatile cyclic fragments in the degradation product is only one third of that estimated based on the content of styrene and the degradation behavior of PS, indicating that the chain transfer reaction is significantly reduced.

The chain breaking of styrene/acrylonitrile (S/AN) copolymer is accelerated, and an monomer is found in the temperature region where volatile degradation products are produced (no monomer is produced when pan is degraded). The proportion of nonvolatile cyclic fragments increased with the increase of an content in the copolymer. Since styrene units prevent the cyclization of propylene cyanogen, some propylene cyanogen units will decompose HCN.

The thermal stability of styrene/vinyl chloride is worse than that of PS. Because of the competitive polymerization rate, the copolymer with VC as the main component cannot be prepared. The thermal stability of the copolymer mainly composed of styrene is worse than that of PS. At about 200 ℃, HCl is released and double bonds are left on the main chain. The double bonds make the main chain of PS more unstable, so the copolymer starts to degrade and release monomers at a lower temperature than pure PS.

7.4 Degradation and stabilization of fluoropolymers

The study of thermal stability of fluoropolymers is different from other polymers, and their thermal behavior is closely related to the environment and test conditions. Specific problems include the effect of oxygen and the tendency of degradation products to react with the instrument, especially the glass parts on the instrument.

7.4.1 Degradation of perfluoropolymers

Polytetrafluoroethylene(PTEE) is one of the most stable polymers. Its heat resistance and medium resistance are excellent, and it is known as the "king of plastics". Its high stability comes from the high strength of C—F bond (487 kJ/mol in CF_4) and the shielding effect of highly negative fluorine atom on the main chain. There is almost no volatilization loss below 450 ℃ in vacuum, and the weight loss is only 30% when the polymer is heated at 500 ℃ for 2 h. In vacuum, compared with polyethylene and other fluoropolymers, the stability order of polymers is as follows: polytetrafluoroethylene > polyvinylidene fluoride > polytrifluoroethylene > polyethylene > polyfluoroethylene > polytrifluorochloroethylene > polyhexafluoropropylene.

The stability of polyhexafluoropropylene (PHFE) is much worse than that of PTFE, and it begins to decompose below 300 ℃. The copolymerization of hexafluoropropylene and tetrafluoroethylene reduces the stability of the polymer.

In the absence of air, 95% of PTFE degradation products are monomers, with a small amount of hexafluoropropylene, but no hydrogen fluoride and molecular chain fragments, unless the temperature is particularly high (1,200 ℃). Although it is not known how the chain breakage is initiated (random chain breakage initiated or structural defects or chain ends), because there is no C—H bond in the polymer molecule and the strength of C—F bond is particularly high, it can be considered that there is no chain transfer reaction in the degradation process of PTFE. Therefore, the degradation mechanism of PTFE is mainly C—C bond homo-cleavage to generate free radicals, followed by free radical depolymerization to generate monomers:

$$\sim\sim\sim CF_2CF_2CF_2CF_2CF_2CF_2CF_2CF_2 \sim\sim\sim \longrightarrow 2 \sim\sim\sim CF_2CF_2CF_2CF_2 \sim\sim\sim$$
$$\sim\sim\sim CF_2CF_2CF_2CF_2 \longrightarrow \sim\sim\sim CF_2CF_2 \sim\sim\sim + CF_2=CF_2$$

Ferry and others studied the degradation process of PTFE under UV irradiation. Pure PTFE and samples containing TiO_2 or CaF_2 were irradiated with high-energy ultraviolet light (laser), and the structure of the samples was analyzed by DSC and optical microscope. The results show that the phase structure of the polymer has a significant effect on the degradation. In pure PTFE, the scattering of microcrystals and the reflection at the wafer beam play a key role in the behavior and degree of polymer degradation. In filled PTFE, the characteristics and content of filler are the most important factors affecting the degradation.

Chapter 7 Degradation and stabilization of different class of commodity polymers

If a filler with absorption function is added, the destruction of the polymer is only limited to the surface. Depending on the filler content, the degradation and destruction of the polymer may change from uneven to uniform distribution. Through further analysis of the propagation path of ultraviolet light in pure PTFE and filled PTFE samples and the scattering and absorption effects of fillers, it is found that the coherent region and scattering region of light have different importance for polymer degradation.

The affinity between PTFE surface and other materials can be improved by proper degradation treatment. Jun and others irradiated the PTFE sample surface with γ-rays in air, then analyzed it with Fourier transform infrared spectroscopy, and measured the contact angle of the sample. The results showed that carbonyl groups were formed on the surface of PTFE after irradiation. As the irradiation dose increases, the number of carbonyl groups increases, the crystallinity of the sample surface decreases, the wettability increases, the dispersion force and the polar part of the surface energy increase, and the friction force increases. When irradiated in air, the dose of 20 kGy is sufficient to improve the adhesion of PTFE surface to most coatings. Tupikov and others studied the degradation of PTFE under far ultraviolet light (123.6 nm). After UV irradiation, the mechanical properties of the samples changed significantly, and the yield stress of the irradiated samples in the stressed state was several times lower than that of the irradiated samples in the unstressed state. During irradiation, the creep rate of the thin film samples increased by more than an order of magnitude. When the thickness of the irradiated thin film was less than 2% of the total thickness, the mechanical properties of PTFE would change.

7.4.2 Degradation of fluorochloropolymers

The substitution of fluorine atom by chlorine atom reduces the stability of polymer and changes the degradation mechanism of polymer, because C—Cl bond is weaker than C—C bond and C—F bond. Therefore, when polytrifluorochloroethylene is degraded in vacuum, volatile degradation products account for 20% of the polymer mass, mainly monomers, as well as some C_3F_5Cl and $C_3F_4Cl_2$. The remaining polymer fragments (CRF) with an average molecular weight of 900.

Since the weakest bond is C—Cl bond, its degradation mechanism may be: first, the C—Cl bond is cracked, and then the chain is broken and depolymerized. In addition, the attack of chlorine atoms on macromolecules may also generate macromolecular free radicals:

$$\sim\sim CF_2CFClCF_2CFClCF_2CFClCF_2CFCl \sim\sim \longrightarrow$$
$$Cl^- + \sim\sim CF_2CFClCF_2CFClCF_2CFCF_2CFCl \sim\sim \longrightarrow$$
$$\sim\sim CF_2CFClCF_2CFCl + CF_2=CFCF_2CFCl \sim\sim$$
$$\sim\sim CF_2CFClCF_2CFCl \longrightarrow \sim\sim CF_2CFCl + \cdot CF_2=CFCl$$

7.4.3 Degradation of fluorohydrogen polymers

Fluorinated polymers containing hydrogen have C–H bonds, which will generate HF when degraded. There are mainly three kinds of such polymers, polyvinylidene fluoride ($-CH_2CHF-$), polyvinylidene fluoride ($-CH_2CF_2-$) and polytrifluoroethylene ($-CF_2CHF-$). The mass percentages of HF yields were about 25%, 50% and 6% respectively when they were degraded in vacuum. The remaining products are mainly CRF. The polymer also changes color due to degradation.

7.4.4 Degradation of other fluoropolymers

Dhal studied the degradation behaviors of three methacrylates containing fluorine atoms in the ester group. These three methacrylates are: poly (2-fluoroethyl) methacrylate (PFEMA), poly (2, 2, 2-trifluoroethyl) methacrylate (PTFEMA), and poly (hexafluoroisopropyl) methacrylate (PHFPMA). The rapid pyrolysis method was used in the experiment. The microgram polymer was cracked at 600 °C in nitrogen atmosphere, and then the volatile products were separated by gas chromatography (GC), and the components were identified by mass spectrometry (MC).

There is no monomer formed during the degradation of PFEMA. Besides the ester group decomposition products, there are also vinyl fluoride, CO_2, acetaldehyde and fluoroacetaldehyde in the products.

7.5 Degradation and stabilization of other vinyl polymers

This section mainly introduces polyvinyl alcohol, polyvinyl acetate, polyacrylonitrile, polyvinyl pyrrolidone and polyvinyl ketone.

7.5.1 Degradation of polyvinyl alcohol and polyvinyl acetate

1. Photo degradation

Most commercial polyvinyl alcohols have strong absorption at 200 - 400 nm, and polyvinyl alcohol samples with low relative molecular weight have stronger absorption than those with high relative molecular weight. This absorption peak is due to the carbonyl groups in the polymer, which are acetaldehyde and oxygen present in the monomer raw materials. Fig. 7.36 shows the UV spectrum of polyvinyl alcohol obtained from polyvinyl acetate polymerized in the presence of acetaldehyde. Mori found that under UV irradiation, carbonyl groups and gel were formed from water-soluble polyvinyl alcohol.

Chain breaking and crosslinking of polyvinyl acetate under ultraviolet (254 nm) irradiation. To explain these reactions, Geuskens and coworkers proposed a mechanism. They assume that the intramolecular hydrogen abstraction reaction of the excited carbonyl group of acetate results in chain breaking. This reaction occurs by forming a transition seven membered ring, which causes the main chain to break but does not remove the side group [see Fig. 7.37 (a)]. They also proposed that the excited acetate group reacts with the hydrogen atom on the adjacent chain [see Fig. 7.37 (b)], and then, the resulting two radicals complex to cause crosslinking.

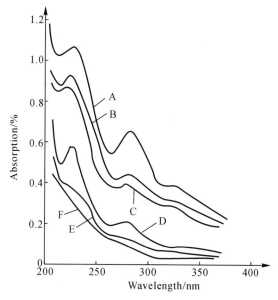

Fig.7.36 UV spectrum of polyvinyl alcohol obtained from polyvinyl acetate polymerized in the presence of acetaldehyde: A 0.48%; B 0.19%; C 0.097%; D 0.02%; E 0.003%; F 0.001%

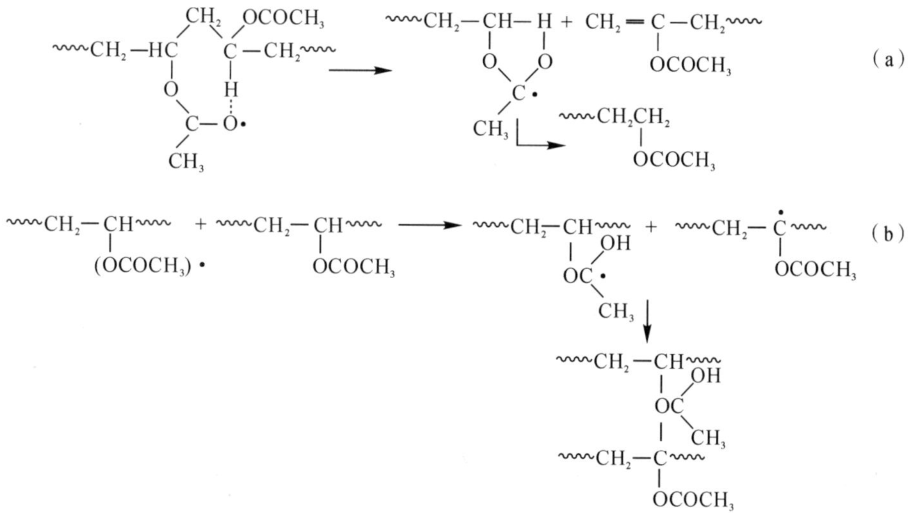

Fig.7.37 Mechanism for the photo degradation of polyvinyl acetate

2. Thermal degradation

Although the thermal stability of polyvinyl acetate is much higher than that of PVC, its degradation characteristics are very similar to that of PVC. When PVAC is degraded, acetic acid is released and the color of polymer becomes dark. In the end, almost every link breaks down into one acetic acid molecule. Like PVC, another stage of degradation occurs at a higher temperature, and the degradation products are also similar to those of PVC at this stage. It was previously thought that the release of acetic acid during the degradation of PVAC was initiated by the chain end, but this view conflicts with the research result that weight loss is independent of relative molecular weight.

When PVAC is programmed in vacuum or nitrogen atmosphere (5 ℃/min or 10 ℃/min), it begins to decompose at nearly 250 ℃, and the maximum rate of degradation in the first stage occurs at 320 – 330 ℃; The maximum rate of degradation in the second stage occurs at about 440 ℃. The carbonization residue at 500 ℃ accounts for about 8% of the initial mass of the polymer. 95% of the degradation products in the first stage are acetic acid, and another 5% are olefins, water, methane, carbon dioxide and carbon monoxide.

Through the study of the degradation behavior of PVAC/PMMA and PVAC/PS alloys, important evidence was found that the degradation of PVAC was carried out according to the free radical mechanism. When the PVAC/PMMA alloy is degraded, the release rate of MMA monomer is greatly accelerated in the range of 250 – 350 ℃, while the release rate of acetic acid is reduced, which is a strong evidence for the existence of free radical mechanism during the degradation of PVAC. When acetic acid is released, a small amount of other products (such as ketones, CO_2, etc.) are considered to be produced by the decomposition of a small amount of acetic acid. In addition, the free radical mechanism can explain the rapid loss of solubility of PVAC during degradation, and a small amount of products other than acetic acid during degradation can also be explained by the side reaction of acetate radical. As shown in Fig. 7.38 is the free radical mechanism of PVAC degradation.

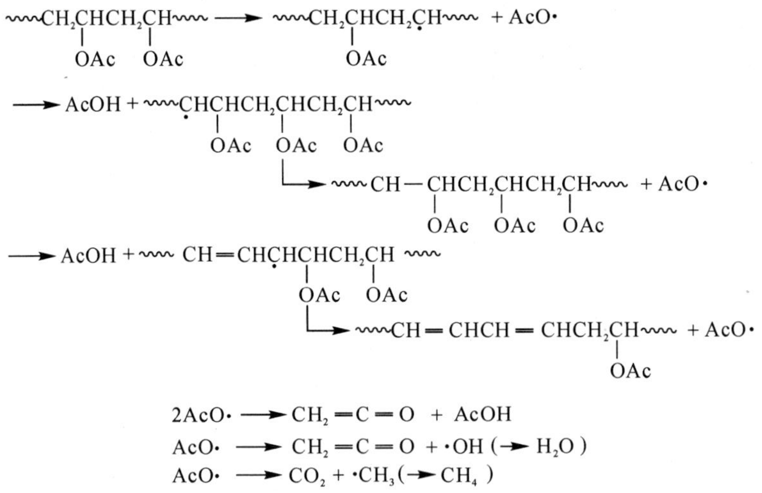

$$\sim\sim CH_2CHCH_2CH\sim\sim \longrightarrow \sim\sim CH_2CHCH_2\overset{\cdot}{C}H\sim\sim + AcO\cdot$$
$$\underset{OAc\ \ OAc}{} \underset{OAc}{}$$

$$\longrightarrow AcOH + \sim\sim \overset{\cdot}{C}HCHCH_2CHCH_2CH\sim\sim$$
$$\underset{OAc\ \ OAc\ \ OAc}{}$$

$$\longrightarrow \sim\sim CH-CHCH_2CHCH_2CH\sim\sim + AcO\cdot$$
$$\underset{OAc\ \ OAc\ \ OAc}{}$$

$$\longrightarrow AcOH + \sim\sim CH=CH\overset{\cdot}{C}HCHCH_2CH\sim\sim$$
$$\underset{OAc\ \ OAc}{}$$

$$\longrightarrow \sim\sim CH=CHCH=CHCH_2CH\sim\sim + AcO\cdot$$
$$\underset{OAc}{}$$

$$2AcO\cdot \longrightarrow CH_2=C=O + AcOH$$
$$AcO\cdot \longrightarrow CH_2=C=O + \cdot OH (\longrightarrow H_2O)$$
$$AcO\cdot \longrightarrow CO_2 + \cdot CH_3 (\longrightarrow CH_4)$$

Fig.7.38 Radical mechanism for the degradation of polyvinyl acetate

Since acetic acid has a catalytic effect on the degradation of PVAC, the molecular mechanism of deacetylation may also exist during the degradation of PVAC.

The thermal degradation of polyvinyl alcohol is very complicate. During degradation, water is decomposed and some conjugated double bonds are formed. Acetaldehyde in the degradation product may be formed by depolymerization of PVA. TVA showed that PVA began to decompose at 200 ℃ under temperature programmed condition.

7.5.2 Degradation of polyacrylonitrile

1. Photo degradation

Polyacrylonitrile (PAN) was irradiated with ultraviolet light in vacuum to produce hydrogen, methane, acrylonitrile and hydrogen cyanide, and chain breaking and cross-linking occurred at the same time. In order to explain some of these reactions, the following mechanism is proposed (see Fig. 7.39): the first step is to generate free radicals, which are generated by main chain breakage or side radical breakage; the next steps include the combination of hydrogen radicals and nitrile radicals to generate hydrogen and hydrogen cyanide, and the depolymerization of macro-molecular radicals to generate monomers.

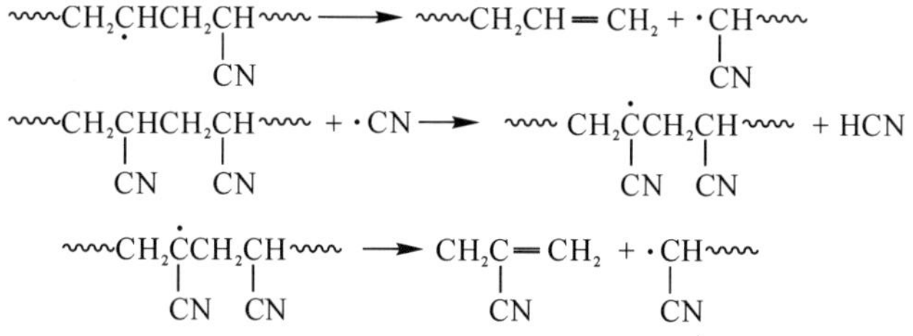

Fig.7.39 Random chain breaking reaction for the degradation of PAN

Jelinek studied the photolysis of polyacrylonitrile in ethylene carbonate $(C_2H_5O)_2CO$ and propylene carbonate $(C_3H_7O)_2CO$. The degradation mechanism of random chain breaks was proposed (see Fig. 7.40). Photooxidation of polypropylene, especially at elevated temperature, cannot lead to degradation, but may form a so-called "trapezoidal structure". The mechanism of this reaction is shown in Fig. 7.41. Due to the formation of partially hydrogenated naphthalene pyridine ring structure during the reaction, the color of the polymer gradually changes from yellow to red and finally to black.

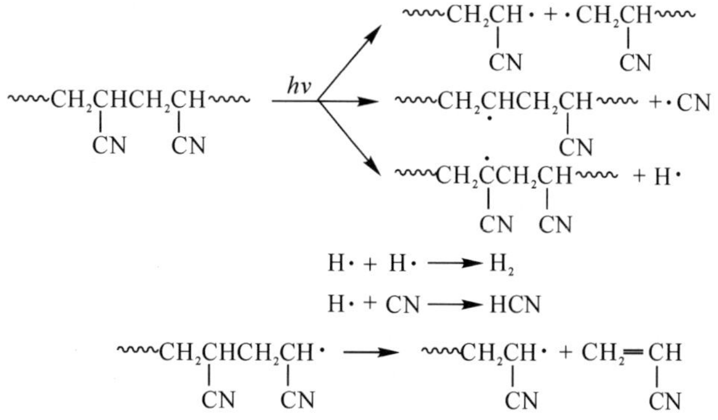

Fig.7.40 Photo degradation of PAN

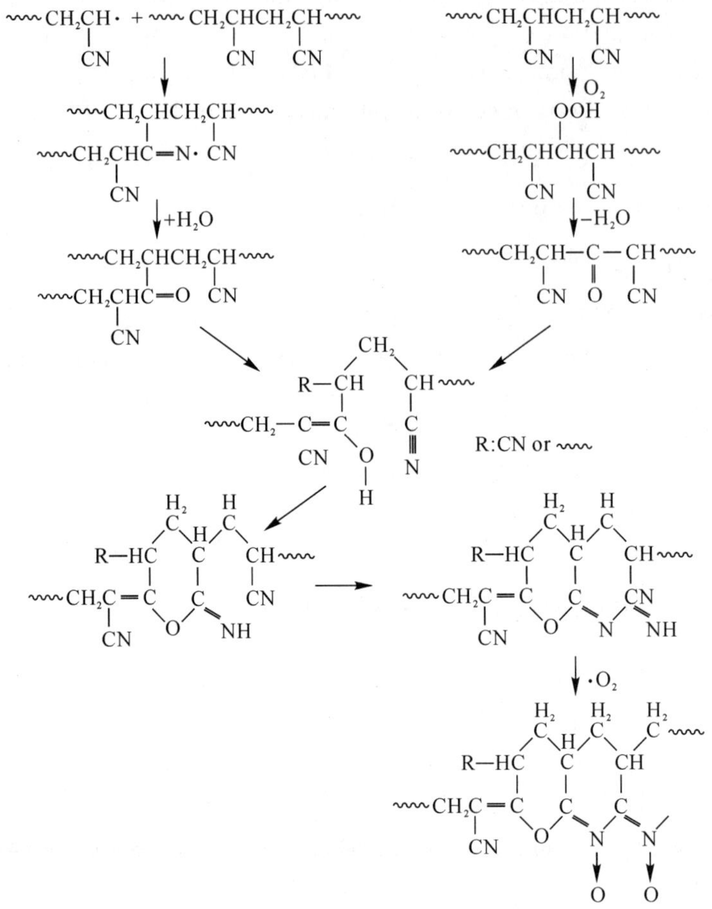

Fig.7.41 Photo-oxidation degradation of PAN

2. Thermal degradation

When polyacrylonitrile is heated, the cyclization occurs at a temperature lower than the temperature required for chain breaking due to the cyclization ability of cyanogens, so no chain breaking occurs. Carbon fiber can be obtained by heating polyacrylonitrile fiber to 100 ℃ under controlled conditions. The color of polyacrylonitrile gradually darkens when heated. When heated in the air at 180 – 190 ℃ for a long time (up to 65 h), it turns yellowish in about 1 h and finally turns brown. If the temperature is programmed in nitrogen or vacuum, the color will quickly turn dark brown at 260 – 270 ℃, and the material will turn black before 350 ℃. The color change at 260 – 270 ℃ corresponds to the exothermic peak on the DTA or DSC curve. The size of the exothermic peak is affected by the size of the polymer sample. The temperature rise of the polymer sample can be more than 40 ℃ higher than that of the environment. Therefore, the TG curve shows a sharp weight loss in the corresponding temperature region, but the weight loss is not large. Combined with TVA curve analysis, the weight loss at this stage is related to the polymer fragments generated by the chain break

Chapter 7 Degradation and stabilization of different class of commodity polymers

caused by the sudden overheating of the polymer. A small amount of volatile components generated at the same time include NH_3 and HCN. TVA curve shows that hydrogen is produced by decomposition above 350 ℃, and nitrogen is produced by decomposition at a high temperature of about 850–900 ℃. Fig. 7.42 for comparison of several thermal analysis curves.

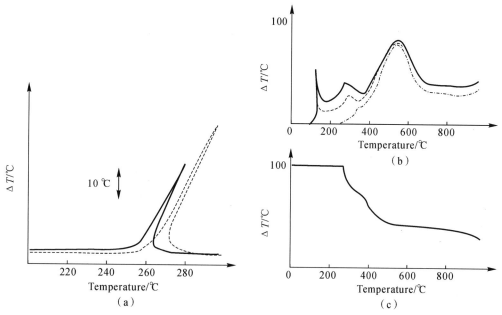

Fig.7.42 Thermal analysis results for the degradation of PAN: (a) DSC under N_2 with heating rate of 10 ℃/min (left) and 5 ℃/min (right); (b) TVA in vacuum with heating rate of 10 ℃/min; (c) TG under N_2 with heating rate of 10 ℃/min
The tempreratre of the cold trap: soidline-0 ℃; dechlhe−75 ℃, chainline−196 ℃

The discoloration of polyacrylonitrile is due to the conversion of cyano group into conjugated (—C=N—) sequence, which is easily triggered by impurities (such as carboxylic acid). After a detailed study of all aspects of this reaction, Grassie concluded that the discoloration of polyacrylonitrile is a free radical process without additives. The exothermic reaction is related to the cyclization of nitrile, which results in the formation of trapezoidal short chain segments. The chain fragment released in the reaction is generated by the fracture of the non cyclized PAN chain segment, and a small amount of HCN is generated by the decomposition of the non cyclized PAN chain segment to remove the side groups. The ammonia produced during degradation comes from the imino terminal group on the ring structure.

The exothermic degradation of pure PAN is unfavorable to the production of carbon fiber, because rapid temperature rise will lead to chain segment fragments. Heat release can be reduced by introducing impurities or comonomers to initiate the cyclization of nitrile, or by controlled preheating in oxygen.

7.5.3 Degradation of polyvinylpyrrolidone

Jellinek studied the photodegradation of polyvinylpyrrolidone in aqueous solution under aerobic conditions. It is found that the photolysis reaction depends on the concentration of polymer in water and is independent of pH. The initiating reaction is usually chain breaking and dehydrogenation. Some monomers are generated in the process of kinetic chain growth reaction. When the temperature increases, more monomers will be produced, which may be due to depolymerization. In addition, chain breakage can also occur through disproportionation reaction. The monomers generated during degradation can be re polymerized. From Fig. 7.43 for main reactions during degradation.

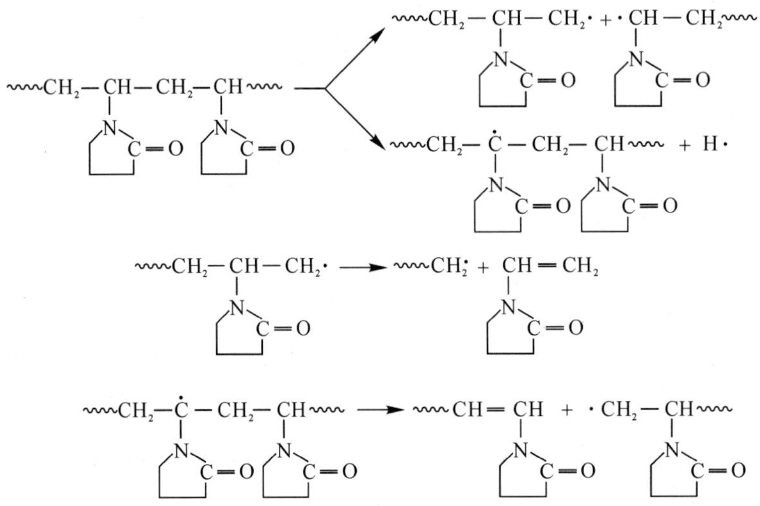

Fig.7.43 Photo degradation of polyvinylpyrrolidone: (a) initiation; (b) depolymerization; (c) disproportionation

7.5.4 Degradation of polyvinyl ketone

Polyvinyl ketones include polymethylvinyl ketone, polyvinylphenyl ketone, polyvinylbenzophenone, polymethylisopropenyl ketone, etc. The common feature of these polymers is that they contain carbonyl groups. They can absorb light with a wavelength above 300 nm. The carbonyl groups are easily excited to the singlet and triplet states. Further light emitting chemical reactions, mainly Norrish I, II and III reactions.

Plooard found that the photolysis of ketone diester homologues $CH_3COO(CH_2)_nCO(CH_2)_nCOOCH_3$ and their copolyesters depends on the n value. when $n = 2$, the main reaction is Norrish I reaction or photoreduction reaction; when $n = 3$, Norrish type II reaction takes place, and its quantum yield is higher than usual; when $n = 4$, both Norrish I reaction and Norish II reaction occur. Their quantum yields are similar to those of the unsubstituted aliphatic ketones under the same photolysis conditions. These authors found that when the ketone group is on the main chain of the polymer, the quantum yield of the broken chain is about 0.01, which seems to be the general rule of the polymer ketone.

Chapter 7 Degradation and stabilization of different class of commodity polymers

At a higher temperature (80 ℃), polymethylvinyl ketone is photolytic according to Norrish Ⅰ reaction. The generated free radicals can also capture hydrogen atoms from the polymer (PH) to generate methane, carbon monoxide and acetaldehyde (see Fig. 7.44), and their quantum yields are 0.000, 6, 0.003 and 0.06 respectively. Guillet and Norrish also measured the total quantum yields of the three compounds at the same temperature (80 ℃), but they believed that the decrease in the relative molecular weight of polymethylvinyl ketone was due to the chain breaking caused by Norrish Ⅱ reaction (see Fig. 7.45).

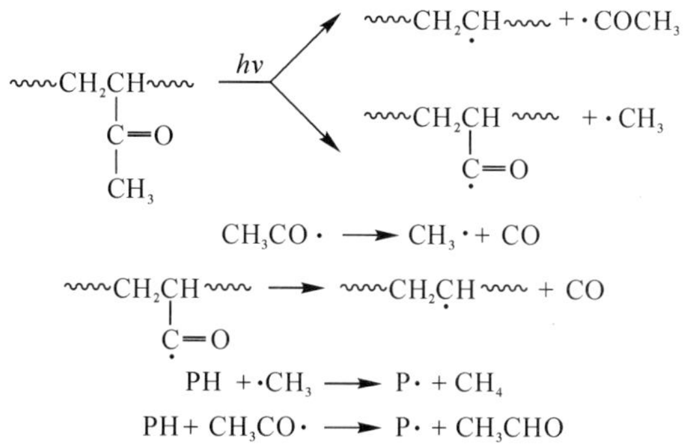

Fig.7.44 Photo degradation of polyvinyl ketone in the form of Norrish Ⅰ reaction

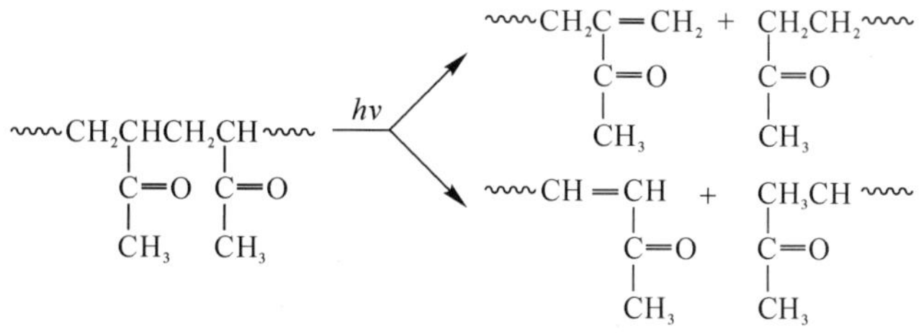

Fig.7.45 Photo degradation of polyvinyl ketone in the form of Norrish Ⅱ reaction

Kato and others studied the photodegradation of the copolymer of polymethylvinyl ketone and methyl methacrylate, and found that the quantum yield of the main chain broken was as high as (0.20 ± 0.02), indicating that the ester carbonyl group hindered the transfer of excitation energy on the main chain. They proposed the following photodegradation mechanism: the chain breaking reaction is carried out through a 7-ring intermediate product, in which hydrogen is taken from methyl group (see Fig. 7.46).

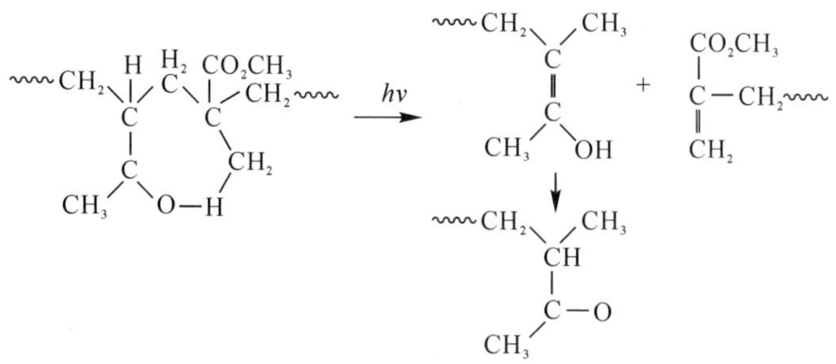

Fig.7.46 Photo degradation of the copolymer of polyvinyl ketone and PMMA

7.6 Degradation and stabilization of acrylate and methacrylate polymers

7.6.1 Degradation of methacrylate polymers

1. Polymethyl methacrylate

The thermal degradation of polymethyl methacrylate (PMMA) is a depolymerization reaction, and the monomer yield is almost 100%. The thermal degradation behavior of PMMA is greatly affected by the polymerization mode in the production of polymer. The polymer obtained by free radical polymerization has unsaturated terminal groups on the molecular chain, which is a weak point that will lead to degradation reaction (see Fig. 7.47).

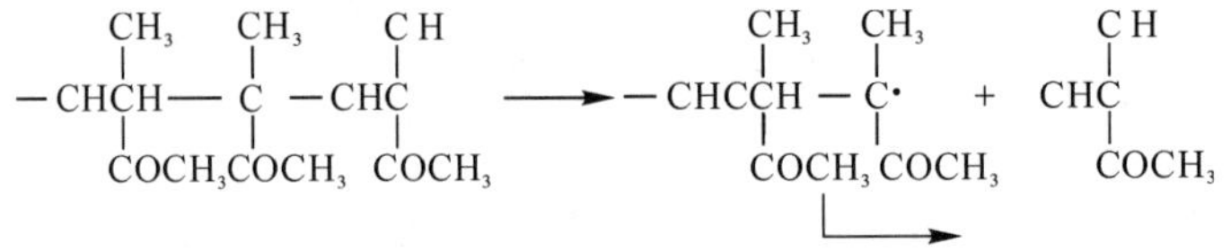

$$-CHCH-\underset{\underset{COCH_3}{|}}{\overset{\overset{CH_3}{|}}{C}}-CHC\underset{\underset{COCH_3}{|}}{\overset{\overset{CH_3}{|}}{}}\overset{\overset{CH}{|}}{\underset{\underset{COCH_3}{|}}{}} \longrightarrow -CHCCH-\underset{\underset{COCH_3}{|}}{\overset{\overset{CH_3}{|}}{C}}\cdot\underset{\underset{COCH_3}{|}}{\overset{\overset{CH_3}{|}}{}} + CHC\overset{\overset{CH}{|}}{\underset{\underset{COCH_3}{|}}{}}$$

Fig.7.47 Depolymerization of PMMA induced by the unsaturated end groups

TVA analysis clearly shows the effect of unsaturated end groups on the thermal degradation of PMMA, as shown in Fig. 7.48. Fig. 7.48 (a) shows the TVA curve of free radical polymerization PMMA. The peak at lower temperature (290 – 300 ℃) represents the degradation reaction caused by unsaturated end groups, and the peak at higher temperature (350 – 400 ℃) represents the degradation reaction caused by the random fracture of the main chain. The size of the two peaks is related to the relative molecular weight of PMMA. For PMMA with very low relative molecular weight, the dynamic chain length of macromolecular free radicals generated by molecular chain splitting exceeds the length of the molecular chain itself. The degradation initiated by unsaturated terminal group can depolymerize this molecular chain completely, so the molecular chain containing unsaturated terminal group will disappear completely in the first stage of degradation. Since the number of molecular chains containing unsaturated end groups in the polymer accounts for about 50%, the size of the two decomposition peaks is roughly equal. For PMMA with higher relative molecular weight, the depolymerization reaction initiated by unsaturated terminal group is not enough to depolymerize the molecular chain completely. Therefore, the random fracture degradation peak of the main chain is larger. The greater the relative molecular weight, the greater the difference between the two peaks.

PMMA synthesized by anionic polymerization has no unsaturated terminal group, and its degradation can only be caused by the random fracture of the main chain, so only one degradation peak appears at a higher temperature, as shown in Fig. 7.48 (b).

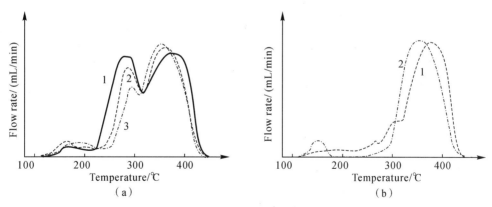

Fig. 7.48 TVA traces for PMMA in vacuum with heating rate of 10℃/min: (a) PMMA from free radical polymerization, line 1: Mn 20,000; line 2: Mn 10,000, line 3: Mn 480,000; (b) PMMA from anionic polymerization, line 1: Mn 60,000; line 2: Mn 1,500,000

Copolymerization of a small amount of other monomers with MMA can prevent the depolymerization of PMMA, thereby improving the stability of PMMA. Some commercial PMMA improves its stability by copolymerizing with a small amount of ethyl acrylate or methacrylic acid. Some small molecules have important influence on the depolymerization of PMMA. Silver acetate can greatly accelerate the depolymerization of PMMA. Zinc bromide changes the depolymerization characteristics of the degradation reaction due to the formation of volatile bromomethane during PMMA degradation. The flame retardant ammonium polyphosphate will react with the polymer to interfere with the reaction of MMA monomer formation.

PMMA does not absorb sunlight, and its light stability is very good. Methyl methacrylate as a component of copolymer can reduce the photo degradation rate of copolymer. The copolymer of styrene and methyl methacrylate, the product of unsaturated polyester cross-linking with methyl methacrylate instead of styrene, has high stability in sunlight.

However, PMMA has a very weak absorption at 254 nm, and there is a weak luminescence phenomenon in the process of illumination. The UV photo degradation of PMMA is temperature dependent. At room temperature, the photo degradation reaction of PMMA includes photo degradation of side ester group, photo degradation of side methyl group and random fracture of polymer main chain (see Fig. 7.49).

At room temperature, the depolymerization reaction caused by each chain break produces about 5 monomer units. Therefore, PMMA only produces relatively few monomers in the low-temperature photolysis reaction. At 159 ℃, the average number of monomer molecules generated by each main chain break increases to 312. When the temperature is higher than 150 ℃, the products of PMMA photodegradation are almost all methyl methacrylate monomers.

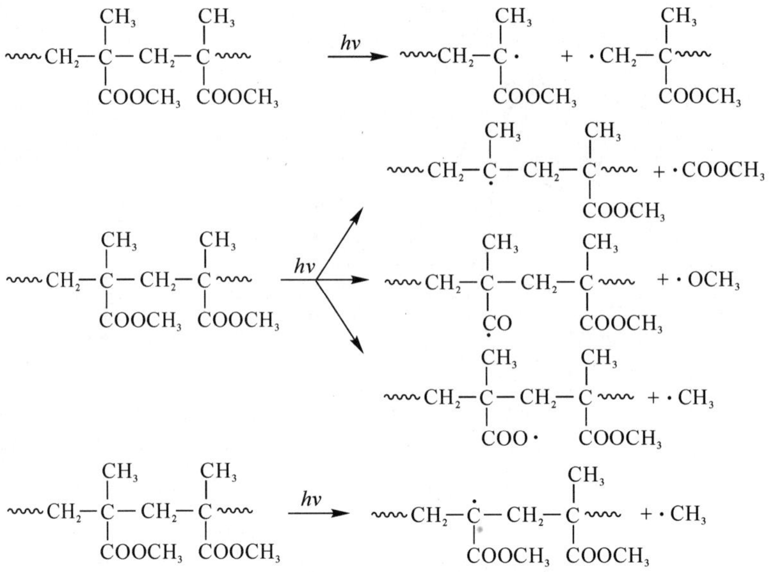

$$\begin{array}{c}\text{~CH}_2\text{-}\underset{\underset{COOCH_3}{|}}{\overset{\overset{CH_3}{|}}{C}}\text{-CH}_2\text{-}\underset{\underset{COOCH_3}{|}}{\overset{\overset{CH_3}{|}}{C}}\text{~} \xrightarrow{hv} \text{~CH}_2\text{-}\underset{\underset{COOCH_3}{|}}{\overset{\overset{CH_3}{|}}{C}}\cdot + \cdot\text{CH}_2\text{-}\underset{\underset{COOCH_3}{|}}{\overset{\overset{CH_3}{|}}{C}}\text{~}\end{array}$$

$$\text{~CH}_2\text{-}\underset{COOCH_3}{\overset{CH_3}{C}}\text{-CH}_2\text{-}\underset{COOCH_3}{\overset{CH_3}{C}}\text{~} \xrightarrow{hv} \begin{cases} \text{~CH}_2\text{-}\underset{\dot{C}}{\overset{CH_3}{C}}\text{-CH}_2\text{-}\underset{COOCH_3}{\overset{CH_3}{C}}\text{~} + \cdot\text{COOCH}_3 \\ \text{~CH}_2\text{-}\underset{CO}{\overset{CH_3}{C}}\text{-CH}_2\text{-}\underset{COOCH_3}{\overset{CH_3}{C}}\text{~} + \cdot\text{OCH}_3 \\ \text{~CH}_2\text{-}\underset{COO\cdot}{\overset{CH_3}{C}}\text{-CH}_2\text{-}\underset{COOCH_3}{\overset{CH_3}{C}}\text{~} + \cdot\text{CH}_3 \end{cases}$$

$$\text{~CH}_2\text{-}\underset{COOCH_3}{\overset{CH_3}{C}}\text{-CH}_2\text{-}\underset{COOCH_3}{\overset{CH_3}{C}}\text{~} \xrightarrow{hv} \text{~CH}_2\text{-}\underset{COOCH_3}{\overset{\cdot}{C}}\text{-CH}_2\text{-}\underset{COOCH_3}{\overset{CH_3}{C}}\text{~} + \cdot\text{CH}_3$$

Fig.7.49 Photo degradation of PMMA

2. Advanced polymethacrylate

Some methacrylates, like polymethyl methacrylate, generate 100% monomers during thermal degradation, while others generate few monomers during thermal degradation.

The degradation behaviors of the four isomers of polybutyl methacrylate are different, and they well represent various thermal degradation situations of polybutyl methacrylate. Polyisobutyl methacrylate produces 100% monomer during pyrolysis; polybutyl methacrylate has a high yield of monomer during pyrolysis, accompanied by a small amount of olefins; polyecbutyl methacrylate has a high yield of olefins with a certain amount of monomers; polytertbutyl methacrylate is mostly olefin, with only a few monomers. McNeil's explanation for the above phenomenon is that the ester group of polymethacrylate will undergo a non free radical decomposition reaction, especially when there is β-atomic hydrogen on alkyl of ester group. When hydrogen atom and ester group can form a six ring transition state, it is particularly easy to decompose into olefins and carboxylic acids. Fig. 7.50 shows the degradation of polysecbutyl methacrylate.

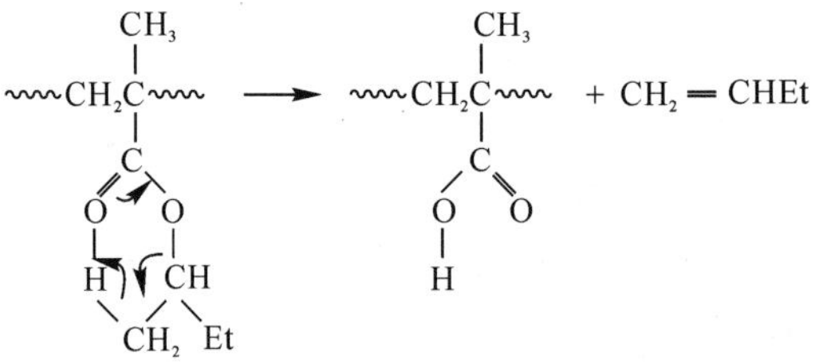

Fig.7.50 Degradation of polysecbutyl methacrylate

Chapter 7 Degradation and stabilization of different class of commodity polymers

The difficulty of ester side group decomposition depends on the β-atomic number of hydrogen. Among the four isomers of polybutyl methacrylate, the atomic number of β-hydrogen is *iso*-butyl ester 1, *n*-butyl ester 2, *sec*-butyl ester 5 and *tert*- butyl ester 9. Therefore, the monomer yield decreases and the olefin yield increases during thermal degradation. The thermal degradation behavior of other polymethacrylates is similar to this.

Both polybutyl methacrylate and poly *tert*-butyl methacrylate become monomers completely under UV irradiation at 170 ℃, but the yield of photo degradation of polybutyl methacrylate depends on the method of polymer preparation. The polymer obtained by bulk polymerization can be degraded completely, while the polymer obtained by solution polymerization can only degrade 40%.

7.6.2 Degradation of acrylate polymer

Although the structure of polyacrylate is close to that of polymethacrylate, its degradation behavior is quite different. When most polyacrylate degrades, its color turns yellow and cross-linking occurs. Many short molecular chain fragments are generated, releasing alcohol and carbon dioxide, and the monomer output is very small (typical value is 0.3% of the total polymer). When there is hydrogen atom on the ester side group, the degradation of polyacrylate also releases olefins, which is the same as that of polymethacrylate.

When methyl polyacrylate is degraded, both cross-linking and chain breaking occur. The main degradation products are molecular chain fragments, only a small amount of volatile products methanol and carbon dioxide, and almost no monomer is produced. There is also no alkyl β-hydrogen atom on the ester side group of benzyl polyacrylate. The main products of its degradation are also molecular chain fragments, as well as a lot of benzyl alcohol and a small amount of CO.

The degradation behaviors of polyethyl acrylate, poly *n*-propyl acrylate, poly *n*-butyl acrylate and polyacrylic acid 2-ethyl hexyl ester are basically similar. Its degradation is completed in one step, and the products are molecular chain fragments, alcohols, CO, and olefins. The yield of molecular chain fragments increases in the above order (15% - 58% mass fraction). Poly isopropyl acrylate (PIPA) and poly *tert*-butyl acrylate (PTBA) show completely different degradation behavior, and their stability is worse. The degradation is divided into two steps, almost every acrylate unit generates an olefin, a small amount of CO and water, and no alcohol. There are 6 and 9 β-alkyl hydrogen atoms on the ester side group of PIPA and PTBA, which are easy to decompose into olefins. Their degradation can be illustrated by Fig. 7.51. The carboxylic acid generated from the initial decomposition dehydrates to form a six circular acid anhydride. Carbon dioxide is generated from the decomposition of —COOH group or acid anhydride group.

It is generally believed that the main degradation mechanism of other polyacrylate is the

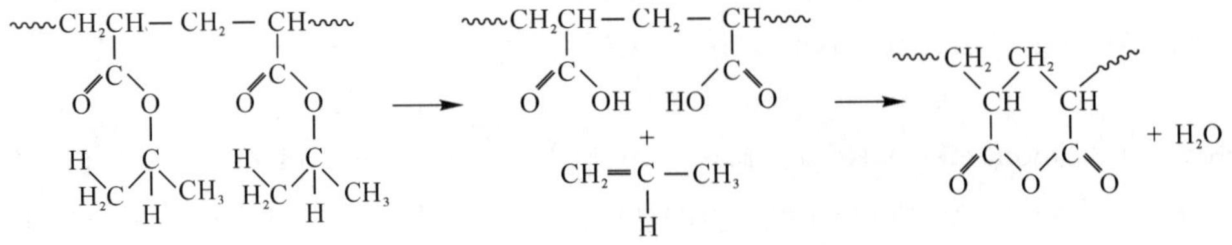

Fig.7.51 The degradation of ester groups in PIPA and PTBA

random fracture of the main chain (see Fig. 7. 52). PMMA, PS and PAMS form two macromolecular free radicals after chain breaking, and one of them undergoes depolymerization reaction to generate monomer. The macromolecular free radicals generated by PMA molecular splitting are different from this. Although the activity of radical B is greater than that of radical A, it can be stabilized by intermolecular transfer of tertiary hydrogen atoms to generate tertiary carbon atom radical C. The free radical can undergo a series of other reactions, including crosslinking and chain breaking. The product of chain breaking reaction is free radical A and a molecular chain containing double bonds. Radical A can also be stabilized by intermolecular transfer of tertiary hydrogen atoms.

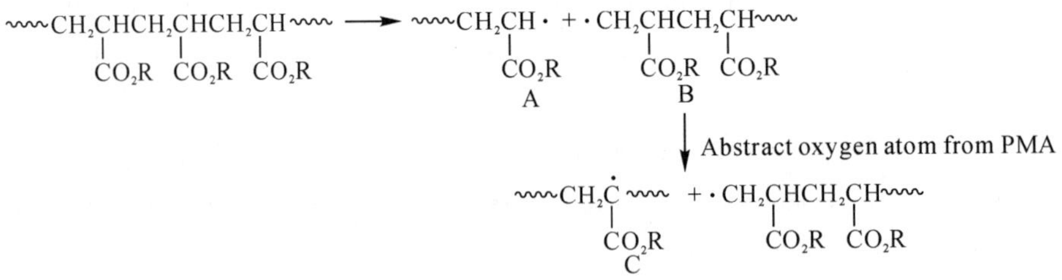

Fig.7.52 Random chain breaking and intermolecular transfer reaction during the degradation of PMA

Some polyacrylate degrades to produce a large amount of alcohol, indicating that there is a reaction along the molecular chain. For example, for PEA, half of the ethyl acrylate unit is decomposed into ethanol, while in the styrene/ethyl acrylate copolymer, the ethyl acrylate unit exists in isolation, and no alcohol is produced when the copolymer is degraded. Free radical A may be the initiator of this reaction, and various mechanisms have been proposed accordingly. Fig. 7. 53 is a possible model.

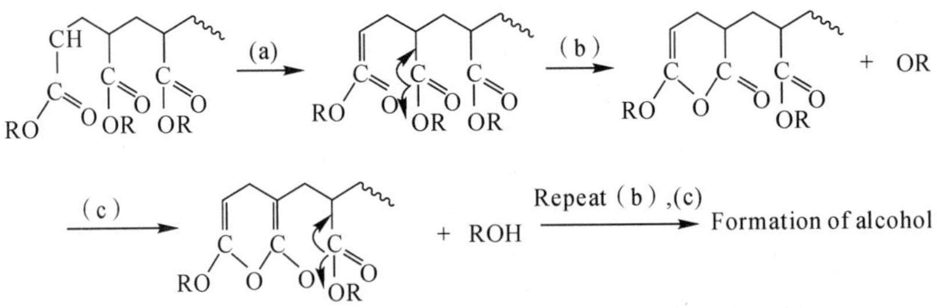

Fig.7.53 Mechanism of the formation of alcohol during the degradation of PMA

Chapter 7 Degradation and stabilization of different class of commodity polymers

Olefins and CO are generated when PEA, PPA, PBA and P_2EHA are degraded. It can be explained by the decomposition of the ester side group of free radical C (see Fig. 7.54). This reaction leads to the formation of products containing double bonds, providing the possibility of further crosslinking.

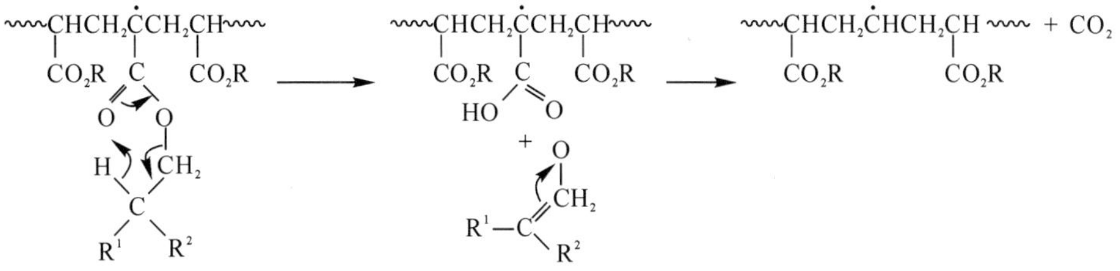

Fig.7.54 Mechanism of the formation of alkene and CO_2 during the degradation of PMA

When polyacrylate is exposed to 254 nm light at room temperature, like PMMA, the C—O bond of ester group breaks, but then the main chain does not break, and cross-linking occurs. When the irradiation temperature decreases, the time required to produce insoluble gel increases. When irradiated in air, chain breaking reaction and cross-linking reaction occur simultaneously, but the presence of oxygen reduces the degree of cross-linking. When the temperature is higher than 150 ℃, the main chain of polyacrylate breaks, but no monomer is generated. When phenyl polyacrylate is irradiated by 310 nm light, molecular rearrangement occurs in both solid and solution states (see Fig. 7.55).

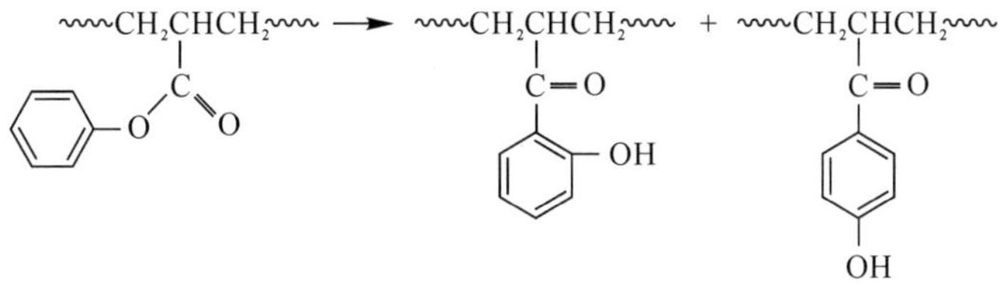

Fig.7.55 Molecular rearrangement during the degradation of phenyl polyacrylate

7.6.3 Degradation of polyacrylic acid and polymethacrylic acid

Only the photo degradation behavior of polyacrylic acid and polymethacrylic acid in aqueous solution has been studied. In the absence of oxygen, the initiation stage reaction of these polymers in direct light is mainly the cleavage of hydrogen connected with tertiary carbon atoms (polyacrylic acid), methyl (polymethacrylic acid) and carboxyl groups.

Chou studied the photo degradation of random and isotactic polymethacrylic acid in aqueous solution, and found that isotactic polymethacrylic acid is a weaker acid than random polymethacrylic acid. The rate of main chain breaking decreases with the increase of pH value. At the same time, the UV spectrum of polymer samples also changes. It is found that the change of chain breaking rate constant with pH value is due to the different entanglement degree of long-chain molecules. The chain breaking constant also increases with the increase of degree of polymerization, which can be explained as the result of intramolecular interaction. A very long molecular chain can absorb more than one photon in a very short time, so synergy can occur. Polymer acids are also very sensitive to the types of electrolytes present in the solution.

The thermal degradation of polymethacrylic acid is carried out in two steps. The first step is to decompose the water at about 200 ℃ to generate cyclic anhydride [see Fig. 7.56 (a)]. At higher temperatures, polyanhydride decomposes to form fragments of various sizes, volatile CO_2, CO and trace olefins, leaving a little carbonization product. The cyclic anhydride structure in Fig. 7.56 (a) plays a very important role in the degradation of polymethacrylic acid, polymethacrylate and its copolymers and alloys. This structure has very typical infrared absorption at 1,805 cm^{-1}, 1,700 cm^{-1} and 1,020 cm^{-1}. When the temperature rises at the rate of 10 ℃/min, it starts to decompose at about 400 ℃, which is slightly higher than the decomposition temperature of the corresponding ester. Since the ring structure can prevent the depolymerization of PMMA molecular chain, the formation of a small amount of anhydride ring on the main chain also has a very obvious role.

The thermal decomposition of polymethacrylamide is similar to that of polymethacrylic acid, and the decomposition is carried out in two steps. The first step is about 200 ℃. Ammonia is mainly released, and a small amount of water is also produced to form imide ring and a small amount of anhydride ring [see Fig. 7.56 (b)]. The second step is to decompose at above 300 ℃ to obtain molecular fragments. At 500 ℃, there are about 20% carbonization residues, obviously more than the residues of polymethacrylic acid.

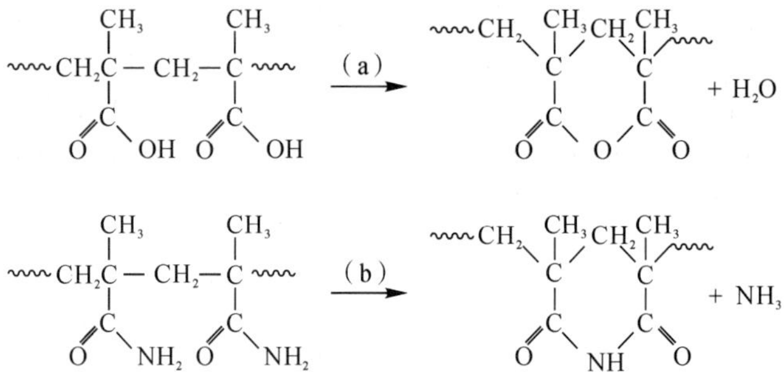

Fig.7.56 Cyclization during the degradation of polymethacrylic acid and polymethacrylamide

A similar reaction occurs during the thermal degradation of polymethacrylic acid (N-methyl) amide, which decomposes methylamine and forms a similar ring structure.

7.6.4 Degradation of polymethacrylate salt

There are not many studies on the degradation of polymethacrylate salt. The literature summarizes the research results of the thermal degradation of polymethacrylate,

polymethacrylate alkali metal salts, alkaline earth metal salts and zinc salts.

Among these polymers, the thermal degradation of polyacrylamide is unique. Its degradation is divided into two steps. The first step is to decompose NH_3 and H_2O and form a ring structure. At higher temperatures, the decomposition produces nonvolatile ring fragment structure compounds, isocyanate, CO_2, CO and CH_4.

The alkali metal salt of polymethacrylic acid starts to decompose at about 300 ℃ under the temperature programmed condition, which is more stable than PMMA. The alkaline earth metal salt of polymethacrylic acid is more stable than that of polymethacrylic acid, and the degradation process is more complex. The degradation products of alkali metal salts of polymethacrylic acid are mainly nonvolatile ring fragment structural compounds, while the degradation products of alkali earth metal salts of polymethacrylic acid have more volatile components. Volatile components always include CO_2 and carbonyl compounds. There is about 6% acetone in the degradation product of alkali metal salt of polymethacrylic acid.

The degradation behavior of polymethacrylate can be shown in Fig. 7.57. The difference of degradation products depends on the competition among various reactions. The stability of the generated carbonate, the volatility of the generated monomer and isobutyrate have important effects. The volatility determines whether the latter can leave the high temperature area quickly without further decomposition.

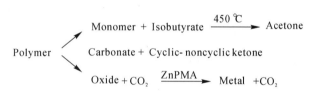

Fig.7.57　Degradation of polymethacrylate salt

7.7　Degradation and stabilization of polydiene polymers

7.7.1　Oxidation and ozonization of polydiene polymers

1. Oxidation of polydiene polymers

Polydiene polymer molecules contain a large number of double bonds, which are either on the main chain (such as 1,4-polymerized polybutadiene and polyisoprene) or on the side group of the polymer (such as 1,2-polybutadiene). Polydiene polymers can be attacked by oxygen even at room temperature, and the reaction is accelerated by light and heat. Therefore, the effect of oxygen and ozone is very important for polydiene polymers.

Pure polydiene polymers have strong absorption only in the far ultraviolet region with wavelength less than 200 nm. However, ordinary samples absorb much longer wavelengths of light up to 240 - 290 nm. Some metal residues in polymers catalyze the decomposition of peroxides to generate free radicals. The oxygen absorption rate of polydiene polymer increases rapidly in the process of illumination and increases with the increase of irradiation time, which is related to the wavelength (see Fig. 7.58).

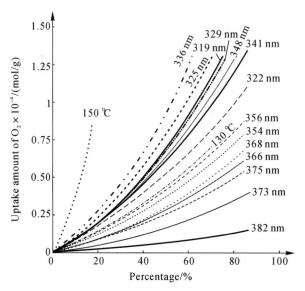

Fig.7.58 Oxygen uptake for natural rubber under different wavelength

Hydroperoxide is produced at the beginning of the photooxidation process. They have strong absorption in near ultraviolet, about 300 – 500 times stronger than polyolefins. After the induction period, the carbonyl group is also formed in the process of ultraviolet irradiation, and its absorption band is about 290 nm. More found that two strong absorption bands appeared at 240 nm and 280 nm in the process of ultraviolet irradiation of polybutadiene, which is believed to be due to the formation of conjugated carbonyl groups.

The photo-oxidation mechanism of polybutadiene polymers can be illustrated by taking cis-1,4-polyisoprene as an example (see Fig. 7.59). The first stage of the reaction is to generate macromolecular alkyl radicals, which are initiated by light, additional alkyl radicals or peroxy radicals [see Fig. 7.59(a)]. Polymeric alkyl radical reacts with oxygen to obtain polymeric peroxy radical [see Fig. 7.59(b)], which can conduct intramolecular reaction and be added to adjacent double bonds to generate cyclic peroxy radical A. Cyclic peroxy radical can react with other oxygen molecules to generate new peroxy radical B [see Fig. 7.59(c)]. The final stage of the oxidation process is the fracture of the polymer cyclic peroxy radical [see Fig. 7.59(d)]. Ultraviolet irradiation, heating or catalysis can initiate fracture reaction. Polymer peroxy radicals react with other polymer molecules to form hydroperoxide [see Fig. 7.59(e)], which can be decomposed into polymer alkoxy radicals and polymer alkyl radicals. The polymer alkoxy radical, the polymer alkyl radical and the peroxide radical can react with each other to crosslink the polymer. It has been confirmed that oxygen bridge is formed during peroxide vulcanization of elastomer.

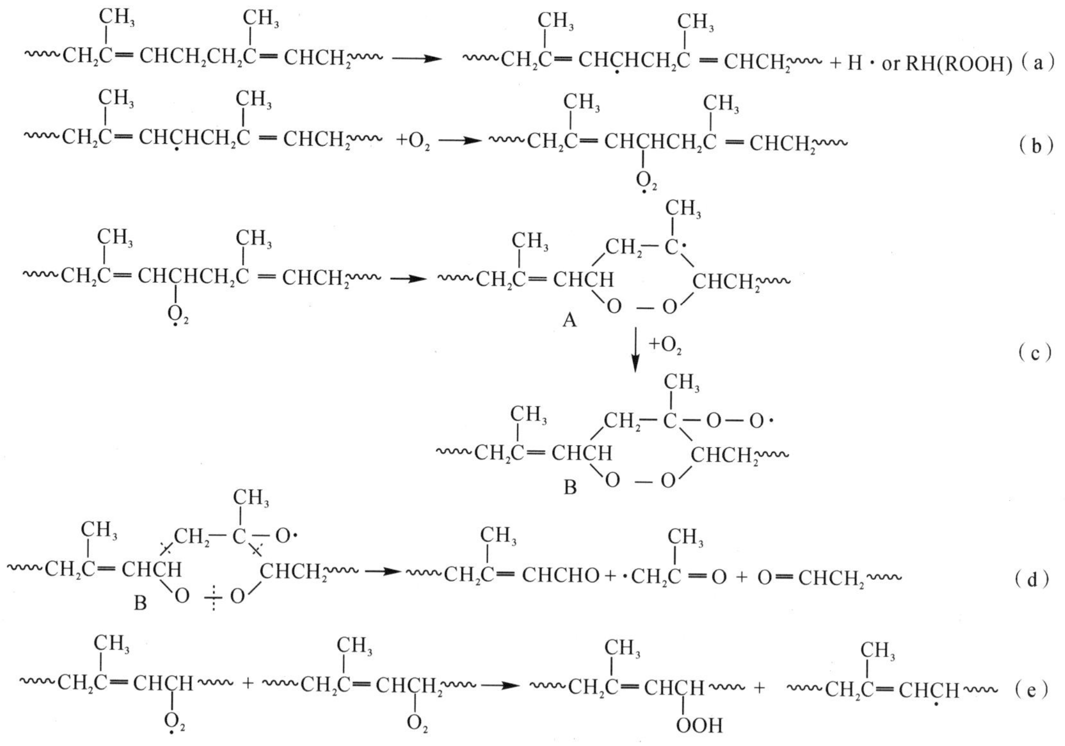

Chapter 7 Degradation and stabilization of different class of commodity polymers

$$\text{\textasciitilde\textasciitilde}CH_2\underset{|}{\overset{CH_3}{C}}=CHCH_2CH_2\underset{|}{\overset{CH_3}{C}}=CHCH_2\text{\textasciitilde\textasciitilde} \longrightarrow \text{\textasciitilde\textasciitilde}CH_2\underset{|}{\overset{CH_3}{C}}=CH\dot{C}HCH_2\underset{|}{\overset{CH_3}{C}}=CHCH_2\text{\textasciitilde\textasciitilde} + H\cdot \text{ or } RH(ROOH) \quad (a)$$

$$\text{\textasciitilde\textasciitilde}CH_2\underset{|}{\overset{CH_3}{C}}=CH\dot{C}HCH_2\underset{|}{\overset{CH_3}{C}}=CHCH_2\text{\textasciitilde\textasciitilde} + O_2 \longrightarrow \text{\textasciitilde\textasciitilde}CH_2\underset{|}{\overset{CH_3}{C}}=CH\underset{|}{\overset{}{C}H}CH_2\underset{|}{\overset{CH_3}{C}}=CHCH_2\text{\textasciitilde\textasciitilde} \quad (b)$$
$$\underset{\dot{O}_2}{}$$

$$\text{\textasciitilde\textasciitilde}CH_2\underset{|}{\overset{CH_3}{C}}=CH\underset{\underset{\dot{O}_2}{|}}{CH}CH_2\underset{|}{\overset{CH_3}{C}}=CHCH_2\text{\textasciitilde\textasciitilde} \longrightarrow \text{\textasciitilde\textasciitilde}CH_2\underset{|}{\overset{CH_3}{C}}=CHCH\underset{O-O}{\overset{CH_2-\overset{CH_3}{\underset{|}{C}}\cdot}{\diagup\diagdown}}CHCH_2\text{\textasciitilde\textasciitilde} \quad$$
$$A$$
$$\downarrow +O_2 \quad (c)$$
$$\text{\textasciitilde\textasciitilde}CH_2\underset{|}{\overset{CH_3}{C}}=CHCH\underset{O-O}{\overset{CH_2-\overset{CH_3}{\underset{|}{C}}-O-O\cdot}{\diagup\diagdown}}CHCH_2\text{\textasciitilde\textasciitilde}$$
$$B$$

$$\text{\textasciitilde\textasciitilde}CH_2\underset{|}{\overset{CH_3}{C}}=CHCH\underset{O\div O}{\overset{CH_2-\overset{CH_3}{\underset{|}{C}}-O\cdot}{\diagup\diagdown}}CHCH_2\text{\textasciitilde\textasciitilde} \longrightarrow \text{\textasciitilde\textasciitilde}CH_2\underset{|}{\overset{CH_3}{C}}=CHCHO + \cdot CH_2\underset{|}{\overset{CH_3}{C}}=O + O=CHCH_2\text{\textasciitilde\textasciitilde} \quad (d)$$
$$B$$

$$\text{\textasciitilde\textasciitilde}CH_2\underset{|}{\overset{CH_3}{C}}=CHCH\text{\textasciitilde\textasciitilde} + \text{\textasciitilde\textasciitilde}CH_2\underset{|}{\overset{CH_3}{C}}=CHCH_2\text{\textasciitilde\textasciitilde} \longrightarrow \text{\textasciitilde\textasciitilde}CH_2\underset{|}{\overset{CH_3}{C}}=CH\dot{C}H\text{\textasciitilde\textasciitilde} + \text{\textasciitilde\textasciitilde}CH_2\underset{|}{\overset{CH_3}{C}}=CHCH\text{\textasciitilde\textasciitilde} \quad (e)$$
$$\underset{\dot{O}_2}{}\underset{\dot{O}_2}{}\underset{OOH}{}$$

Fig. 7.59 Degradation mechanism for polydienes

2. Ozonization of polydiene polymers

The reaction between ozone and olefin compounds (including polydiene polymers) is generally believed to be the addition reaction between oxygen and double bonds. The process is roughly as follows (Fig. 7.60): (a) under the influence of ozone molecules, the π bond of the double bond is polarized, and ozone molecules interact with the double bond to form molecular ozone compounds; (b) molecular ozone compounds undergo isomerization to generate isoozone compounds; (c) isoozone compounds are extremely unstable and break quickly to form double charged peroxides and ketones. It is still unclear in what form the ozone compounds generated continue to split.

7.7.2 *Cis*-1,4-polybutadiene

When *cis*-or *trans*-1,4-polybutadiene is irradiated with 254 nm ultraviolet light in vacuum, they are isomerized. For this reaction, the following reaction mechanism is proposed: the electrons on the double bond are excited to the anti bond state, where they can rotate freely, so mutual transformation of spatial configurations can occur (see Fig. 7.61).

The experiment confirmed that in the process of irradiation of *cis*-1,4-polybutadiene containing ($-CH_2-CH=CH-CH_2-$) unit with ultraviolet light, there was no double bond migration on the main chain, and only *cis-trans* isomerization occurred. The reduction

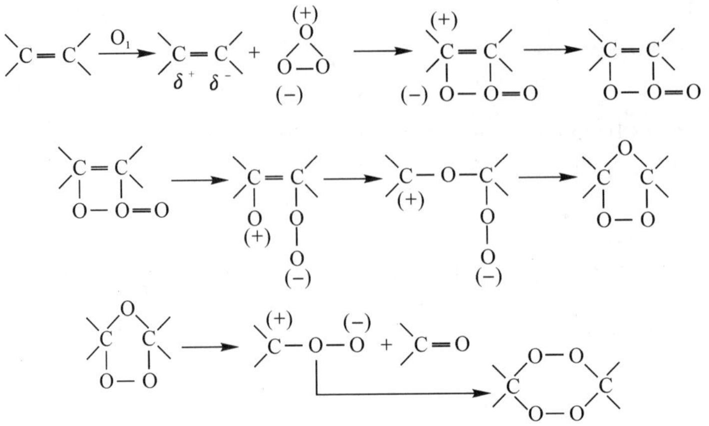

Fig.7.60 Interaction between ozone and alkene

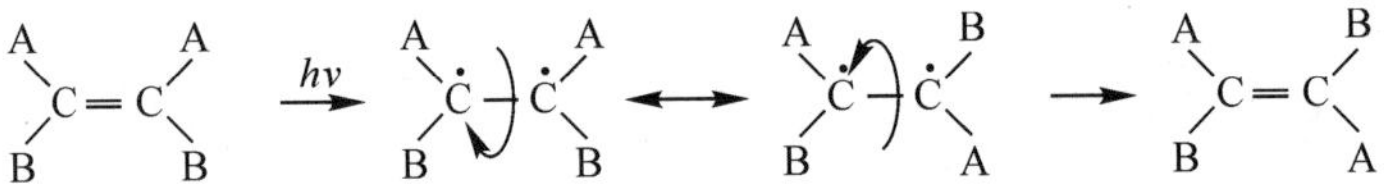

Fig.7.61 Photo isomerization of 1,4-polybutadiene.

of double bonds during the reaction can be explained by the cyclization reaction. The quantum yield of cis and trans isomerization and double bond disappearance is one order of magnitude greater when irradiated with 124 nm UV light than when irradiated with 254 nm UV light. The weak bond of polydiene chain is CH—CH bond (230 kJ/mol), and the bond energy is lower than that of common C—C bond (347 kJ/mol).

This is due to the resonance energy of two allyl radicals generated when the main chain breaks:

$$\sim\sim CH_2-CH=CH-CH_2-CH_2-CH=CH-CH_2\sim\sim \longrightarrow$$
$$\sim\sim CH_2-CH=CH-CH_2\cdot + \cdot CH_2-CH=CH-CH_2\sim\sim$$

The breakage of double bond main chain may be accompanied by the migration of alkenyl group to the end:

$$\sim\sim CH_2-CH=CH-CH_2\cdot \longrightarrow \sim\sim CH_2-\overset{\cdot}{C}H-CH=CH_2$$

Cyclopropyl may also be generated according to the following mechanism:

$$\sim\sim CH_2-CH=CH-CH_2\sim\sim \xrightarrow{h\nu} \sim\sim CH_2-\overset{\cdot}{C}H-\overset{\cdot}{C}H-CH_2\sim\sim \longrightarrow$$
$$\sim\sim CH_2-\overset{\cdot}{C}H-CH_2-\overset{\cdot}{C}H\sim\sim \longrightarrow$$

7.7.3 Polypentaries

Compared with polybutadiene, the polymers obtained by head tail polymerization of pentadiene have one methylene group in each monomer unit. This leads to the enhancement of $CH_2—CH_2$ bond at the "allyl position". In the stereoregular polypentadiene, one such

Chapter 7 Degradation and stabilization of different class of commodity polymers

bond breaking generates one allyl radical and one alkyl radical, while the bond breaking of 1,4-polybutadiene generates two allyl radicals. The bond strength of each CH_2-CH_2 in polypentadiene is 291 kJ/mol.

Polypentadiene is similar to polybutadiene, showing photo induced *cis-trans* isomerization, but less degradation. Chain breaking reaction is carried out according to the following mechanism.

$$\sim\sim CH_2-CH=CH-CH_2-CH_2-CH_2-CH=CH-CH_2-CH_2 \sim\sim \xrightarrow{h\nu}$$
$$\sim\sim CH_2-CH=CH-CH_2-CH_2 \cdot + \cdot CH_2-CH=CH-CH_2-CH_2 \sim\sim$$

Double bond migration can be carried out on the generated allyl radical, and no cyclopropyl structure is found.

7.7.4 1,2-polybutadiene

The double bond of 1,2-polybutadiene is on the side group of the main chain. In the process of ultraviolet irradiation, there is no chain breaking of 1,2-polybutadiene, only cyclization reaction occurs, forming cyclohexane ring structure [see Fig. 7.62(a)].

Photo-rearrangement of 1,2-polybutadiene can also generate other structures [see Fig. 7.62(b)].

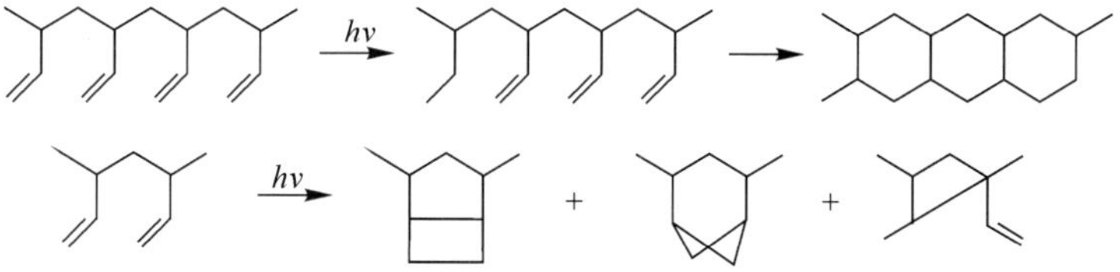

Fig.7.62 Photo rearrangement of 1,2-polybutadiene

7.7.5 *Cis*-1,4-polyisoprene

The weakest bond in polyisoprene is the bond between two methylene groups (C_4-C_5). After the bond breaks, two free radicals A and B with allyl structure are generated. Double bond migration can occur in free radicals A and B. All free radicals (A, B, C and D) can be compounded or added to the double bonds of their own chains or other polymer chains, as shown in Fig. 7.63(d) and (e).

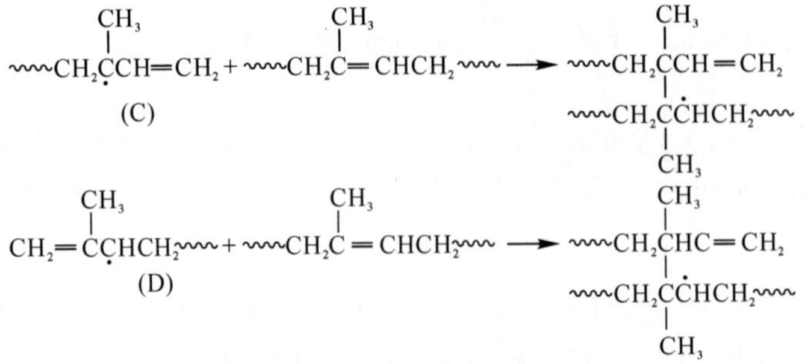

$$\text{~CH}_2\dot{\text{C}}(\text{CH}_3)\text{CH}=\text{CH}_2 + \text{~CH}_2\text{C}(\text{CH}_3)=\text{CHCH}_2\text{~} \longrightarrow \text{~CH}_2\text{C}(\text{CH}_3)\text{CH}=\text{CH}_2 \;/\; \text{~CH}_2\text{C}(\text{CH}_3)\dot{\text{C}}\text{HCH}_2\text{~} \quad (d)$$
(C)

$$\text{CH}_2=\text{C}(\text{CH}_3)\dot{\text{C}}\text{HCH}_2\text{~} + \text{~CH}_2\text{C}(\text{CH}_3)=\text{CHCH}_2\text{~} \longrightarrow \text{~CH}_2\text{C}(\text{CH}_3)\text{HC}=\text{CH}_2 \;/\; \text{~CH}_2\text{C}(\text{CH}_3)\dot{\text{C}}\text{HCH}_2\text{~} \quad (e)$$
(D)

Fig.7.63 Photo degradation of polyisoprene

Hydrogen capture reaction occurs when *cis*-1,4-polyisoprene is irradiated with 230 – 360 nm ultraviolet light. The proposed reaction mechanism is shown in Fig. 7.64. The macromolecular free radicals combine to form cross-linking.

The photoexcitation of double bonds leads to *cis-trans*-isomerization and the formation of cyclopropane groups. The cyclopropane type structure is produced by the migration reaction of double radicals through 1,2-hydrogen, as shown in Formula (a) in Fig. 7.65. According to the electron paramagnetic resonance study, another mechanism for the formation of cyclopropyl is also proposed, as shown in Fig. 7.65(b) and (c).

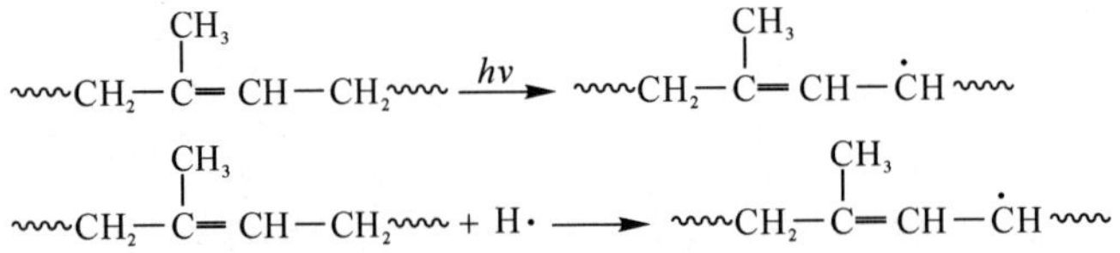

$$\text{~CH}_2-\text{C}(\text{CH}_3)=\text{CH}-\text{CH}_2\text{~} \xrightarrow{h\nu} \text{~CH}_2-\text{C}(\text{CH}_3)=\text{CH}-\dot{\text{C}}\text{H}\text{~}$$

$$\text{~CH}_2-\text{C}(\text{CH}_3)=\text{CH}-\text{CH}_2\text{~} + \text{H}\cdot \longrightarrow \text{~CH}_2-\text{C}(\text{CH}_3)=\text{CH}-\dot{\text{C}}\text{H}\text{~}$$

Fig.7.64 Hydrogen abstraction reaction during the photo degradation of polyisoprene

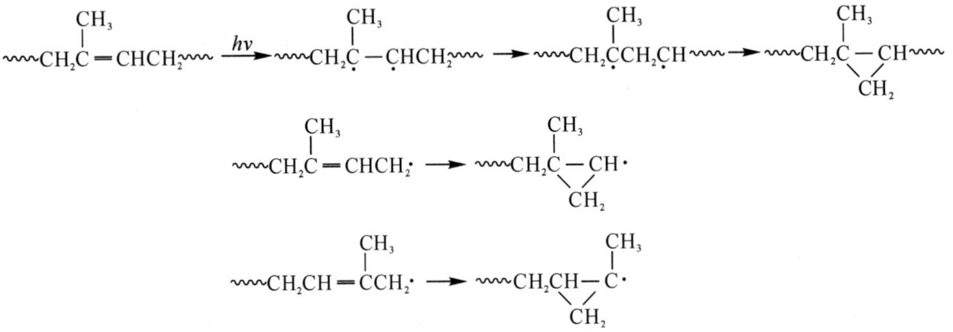

$$\text{~CH}_2\text{C}(\text{CH}_3)=\text{CHCH}_2\text{~} \xrightarrow{h\nu} \text{~CH}_2\dot{\text{C}}(\text{CH}_3)-\dot{\text{C}}\text{HCH}_2\text{~} \longrightarrow \text{~CH}_2\dot{\text{C}}(\text{CH}_3)\text{CH}_2\dot{\text{C}}\text{H}\text{~} \longrightarrow \text{~CH}_2\text{C}(\text{CH}_3)-\text{CH}\text{~} \; (\triangle\text{CH}_2) \quad (a)$$

$$\text{~CH}_2\text{C}(\text{CH}_3)=\text{CHCH}_2\cdot \longrightarrow \text{~CH}_2\text{C}(\text{CH}_3)-\dot{\text{C}}\text{H}\; (\triangle\text{CH}_2) \quad (b)$$

$$\text{~CH}_2\text{CH}=\text{C}(\text{CH}_3)\text{CH}_2\cdot \longrightarrow \text{~CH}_2\text{CH}-\dot{\text{C}}(\text{CH}_3)\; (\triangle\text{CH}_2) \quad (c)$$

Fig.7.65 Cyclopropyl formation reaction during the photo degradation of polyisoprene

Chapter 7. Degradation and stabilization of different class of commodity polymers

Further reading I : thermal oxidative degradation of polypropylene

Isotactic polypropylene (i-PP) is a major commodity plastic material which cannot be utilized without thermal stabilizers. With a moderately complex structure, i-PP is frequently used as a "model system" to test the different theoretical and experimental approaches to macromolecular degradation.

1. Initiation

The homogeneous oxidation of PP follows the free-radical auto-oxidation mechanism. Under isothermal conditions, the oxygen uptake curves display a pseudo-induction period during which oxidation is auto accelerated. The duration for the induction period t_i depends on the sample purity and preoxidation history. For clean samples, t_i is reasonably reproducible and is of the same order of magnitude as the POOH lifetime, determined by iodometric titration. The temperature dependence of the induction time obeys an Arrhenius law with apparent activation energy of 105 ± 15 kJ/mol, which is the same as for the decomposition of hydroperoxides. The corresponding rate constants are much lower than for the other degradation processes and account for the induction period during which hydroperoxides accumulate before reaching a maximum. It has long been recognized that residual Ziegler-Natta polymerization catalysts, generally at the 1 - 20 ppm level, accelerate the solid-state degradation of PP. The negative influence of polymerization catalyst residues depends not only on the type of catalyst, but also on its concentration. The induction period stage is complicated by the morphology of the sample. In i-PP, for instance, a sample with small spherulite sizes ($<$ 100 mm), obtained by rapid quenching, has a short induction time (see Fig. 7.66). A large spherulite sample (350 - 500 mm), obtained by prolonged annealing, results in a significant increase in the induction period. The difference was explained by a difference in the oxygen diffusion rate in the oriented amorphous regions, which are more strained in large spherulite structures.

It was shown from decomposition kinetics and by treatment with dimethyl sulfide that peroxides consist of two types: a fast-decomposing one composed of peracids, and a slowly decomposing one consisting of hydroperoxides and hydroperesters. During the induction period, the slowly decomposing hydroperoxides accumulate and the oxidation rate is controlled by the rate of decomposition, which may be finally catalyzed by metal ion residues. The auto acceleration stage is controlled by the fast-decomposing peracids.

Oxygen uptake (left-hand scale) and hydroperoxide formation (right-hand scale) in the thermo-oxidative degradation of i-PP films at 130 ℃, with small (—) and large (- -- -) spherulites.

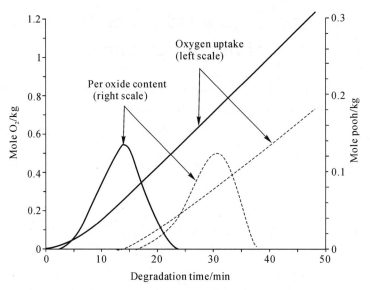

Fig.7.66 Degradation of *i*-pp under different conditions

Regardless of the exact mechanism of bond scission, the mechanistic step for initiation in PP can be described schematically by following equation:

$$\sim\sim CH_2-\underset{\underset{CH_3}{|}}{\overset{\cdot}{C}H}-CH_2\sim\sim \xrightarrow{+X\cdot} \begin{cases} \sim\sim \overset{\cdot}{C}H-\underset{\underset{CH_3}{|}}{CH}-CH_2\sim\sim \\ \sim\sim CH_2-\underset{\underset{CH_3}{|}}{\overset{\cdot}{C}}-CH_2\sim\sim \end{cases}$$

2. Propagation

For PP, the radical conversion step with oxygen is given by following equation:

$$\sim\sim CH_2-\underset{\underset{\cdot}{|}}{\overset{CH_3}{C}}-CH_2-\underset{\underset{H}{|}}{\overset{CH_3}{C}}\sim\sim \xrightarrow{+O_2} \sim\sim CH_2-\underset{\underset{O-O\cdot}{|}}{\overset{CH_3}{C}}-CH_2-\underset{\underset{H}{|}}{\overset{CH_3}{C}}\sim\sim$$

Hydrogen abstraction from tertiary carbon has an activation energy lower by approximately 15 kJ/mol than an abstraction reaction on secondary carbon, and is the predominant mode of formation of hydroperoxides in PP. In addition to energy considerations, the hydrogen abstraction rate constant is dependent on steric factors and polymer conformation. It is found, for instance, that k_3 in solution is lower in a theta solvent than in a good solvent, owing to increased steric repulsion in a contracted molecular coil.

Abstraction reaction in polymers can occur intramolecularly or intermolecularly. The former possibility is important in polymers which possess lateral groups (PP and PS), whereas it is nonexistent in linear polymers (HDPE). The possibility of intramolecular H abstraction has been advanced as one of the reasons for the high sensitivity of PP (in comparison to HDPE) toward oxidative degradation. Infrared studies have revealed that

Chapter 7 Degradation and stabilization of different class of commodity polymers

more than 90% of hydroperoxides in PP are hydrogen-bonded in sequences of two or more. This result supports an intramolecular hydrogen abstraction reaction, facilitated by a six-membered ring stereochemical arrangement (following equation).

$$\sim\sim CH_2-\underset{\underset{O\diagdown O\cdot}{|}}{\overset{\overset{CH_3}{|}}{C}}-CH_2-\underset{H}{\overset{\overset{CH_3}{|}}{C}}\sim\sim \longrightarrow \sim\sim CH_2-\underset{\underset{O\diagdown OH}{|}}{\overset{\overset{CH_3}{|}}{C}}-CH_2-\overset{\overset{CH_3}{|}}{\underset{\cdot}{C}}\sim\sim$$

$$\xrightarrow{+O_2} \sim\sim CH_2-\underset{\underset{O\diagdown OH}{|}}{\overset{\overset{CH_3}{|}}{C}}-CH_2-\underset{\underset{O\diagdown O\cdot}{|}}{\overset{\overset{CH_3}{|}}{C}}-CH_2-\underset{H}{\overset{\overset{CH_3}{|}}{C}}\sim\sim$$

3. Chain branching

Chain branching by homolytic decomposition of polymer hydroperoxides results in formation of highly reactive polymer alkoxy (PO·) and hydroxyl (·OH) radicals. The small and mobile ·OH can readily abstract hydrogen from a nearby polymer chain, creating a secondary or tertiary macroalkyl radical. The tertiary alkoxy radical can abstract hydrogen intramolecularly, forming a primary alkyl radical as in following equation:

$$\underset{H}{\overset{CH_3\diagdown \diagup CH_2\sim\sim}{\underset{\underset{O\cdot}{|}}{C}}}\overset{|}{\underset{CH_2}{\overset{CH\sim\sim}{|}}} \longrightarrow \underset{H}{\overset{CH_3\diagdown \diagup CH_2\sim\sim}{\underset{\underset{O}{|}}{C}}}\overset{|}{\underset{\cdot CH_2}{\overset{CH\sim\sim}{|}}}$$

The polyalkoxy radicals can decompose further by b-scission, yielding ketones, aldehydes, and alkyl radicals, depending on the initial position of the radical (following equations).

$$\sim\sim CH_2-\underset{\underset{O\cdot}{|}}{\overset{\overset{CH_3}{|}}{C}}-CH_2-\underset{H}{\overset{\overset{CH_3}{|}}{CH}}\sim\sim \xrightarrow{\text{-Scission}} \begin{array}{l} \sim\sim CH_2-\underset{\underset{O}{\|}}{\overset{\overset{CH_3}{|}}{C}} + \cdot CH_2-\overset{\overset{CH_3}{|}}{CH}\sim\sim \\[6pt] \sim\sim CH_2-\underset{\underset{O}{\|}}{\overset{\overset{CH_3}{|}}{C}}-CH_2-\overset{\overset{CH_3}{|}}{CH}\sim\sim + \cdot CH_3 \end{array}$$

$$\sim\sim CH-\underset{\underset{O\cdot}{|}}{\overset{\overset{CH_3}{|}}{CH}}-CH\sim\sim \xrightarrow{\text{-Scission}} \sim\sim CH-\underset{\underset{O}{\|}}{\overset{\overset{CH_3}{|}}{C}}\overset{H}{\diagup} + \cdot CH\sim\sim$$

Similarly, the cleavage of peroxy radicals results in the formation of double bonds along with aldehydes and ketones (following equations).

$$\sim\sim CH-\underset{\underset{O\diagdown O\cdot}{|}}{\overset{\overset{CH_3}{|}}{CH}}-CH\sim\sim \xrightarrow{\text{-Scission}} \sim\sim CH-\underset{\underset{O}{\|}}{\overset{\overset{CH_3}{|}}{C}}\overset{H}{\diagup} + HC=CH\sim\sim + \cdot OH$$

$$\sim\sim CH_2-\underset{\underset{O\diagdown O\cdot}{|}}{\overset{\overset{CH_3}{|}}{C}}-CH_2-\overset{\overset{CH_3}{|}}{CH}\sim\sim \xrightarrow{\text{-Scission}} \sim\sim CH_2-\underset{\underset{O}{\|}}{\overset{\overset{CH_3}{|}}{C}} + CH_2=\overset{\overset{CH_3}{|}}{CH}\sim\sim + \cdot OH$$

4. Termination

In an oxygen-deficient atmosphere, vinylidene and vinyl compounds may be formed from the disproportionation of polypropylene radicals (following equations).

$$2P-\underset{\underset{CH_3}{|}}{CH}-CH_2 \cdot \rightarrow P-\underset{\underset{CH_3}{|}}{C}=CH_2 + P-\underset{\underset{CH_3}{|}}{CH}-CH_3$$

$$2P-CH_2-CH \cdot \rightarrow P-CH_2-CH=CH_2 + P-CH_2-CH_2-CH_3$$

In the presence of oxygen, the majority of recombinations occur between tertiary peroxy radicals, to give dialkyl peroxides and oxygen (following equation).

$$\sim\sim CH_2-\underset{\underset{O-O \cdot}{|}}{\overset{\overset{CH_3}{|}}{C}}-CH_2\sim\sim + \sim\sim CH_2-\underset{\underset{\cdot O-O}{|}}{\overset{\overset{CH_3}{|}}{C}}-CH_2\sim\sim \longrightarrow$$

$$\sim\sim CH_2-\underset{\underset{O-O}{|}}{\overset{\overset{CH_3}{|}}{C}}\sim\sim\sim\sim\underset{\underset{}{|}}{\overset{\overset{CH_3}{|}}{C}}-CH_2\sim\sim \longrightarrow \sim\sim CH_2-\underset{\underset{O-O}{|}}{\overset{\overset{CH_3}{|}}{C}}\sim\sim\sim\sim\underset{\underset{}{|}}{\overset{\overset{CH_3}{|}}{C}}-CH_2\sim\sim + O_2$$

$$H = -300 \text{ kJ/mol}$$

Further reading II : PVC heat stabilizers

In spite of its inherent thermal instability, PVC possesses several attractive properties, such as economy of production and processing, and the ease of variation of its properties, with appropriate blending with other polymers or additives, from hard and tough materials to elastomeric ones. Currently, PVC ranks second only to polyolefins in terms of worldwide production among industrial polymers. This technical success is the result of considerable research in the field of degradation and stabilization of this polymer since its introduction as a commodity plastic in the middle of the 20th century.

The low thermal stability of PVC originates from the presence of labile structures and of the autocatalytic deleterious effect of the hydrochloric acid evolved. Thermal stabilizers for PVC consist principally of metal carboxylates and organotin compounds (primary stabilizers), used in combination for preventive and curative functions. As with degradation, uncertainties continue to exist in the exact stabilizing mechanisms of these additives. There is evidence that organotin derivatives stabilize PVC by substituting the labile allylic chorine with a more thermally stable thioether group (following equation).

$$(C_4H_9)_2Sn(SCH_2COO-C_8H_{17})_2 + \sim\sim CH_2-CH=CH-CHCl-CH_2\sim\sim$$
$$\rightarrow \sim\sim CH_2-CH=CH-CH(SCH_2COO-C_8H_{17})-CH_2\sim\sim$$
$$+ Cl-Sn(C_4H_9)_2(SCH_2COO-C_8H_{17})$$

Hydrogen chloride is bound with metallic (Zn, Ca, Ba) organic acid salts with formation of the metal chloride and the corresponding free fatty acids. The formation of polyene sequences can be prevented by combination reactions with thiol or maleate derivatives. Although efficient in blocking the degradation, most of the inorganic stabilizers leave toxic residues and current research is focused on the development of new, less polluting, organic stabilizers.

Appendix Ⅰ Typical wordlist used in this book

In this section, a wordlist for polymer degradation is given as following to help the readers to further understand the English edition of "Polymer Degradation".

Typical wordlist used in this book

No.	English	Chinese	No.	English	Chinese
1	Aging	老化	28	Crystalline	结晶度
2	Degradation	降解	29	Amorphous	无定形的
3	Deterioration	劣化	30	Supramolecular	超分子的
4	Resin	树脂	31	Aggregation	聚集态
5	Cracking	龟裂	32	Impurity	杂质
6	Aromatic	芳香的	33	Auxiliary	辅助的
7	Polyamides	聚酰胺	34	Monomer	单体
8	Thermoplastic	热塑性	35	Emulsifier	乳化剂
9	Processability	可加工性	36	Dispersant	分散剂
10	Thermoset	热固性	37	Stabilizer	稳定剂
11	Polypropylene	聚丙烯	38	Molding	成型
12	Crosslink	交联	39	Dye	染料
13	Intrinsic cause	内因	40	Pigment	颜料
14	Extrinsic cause	外因	41	Plasticizer	增塑剂
15	Polystyrene	聚苯乙烯	42	Filler	填料
16	Poly tetra fluoroethylene	聚四氟乙烯	43	Peroxyl radicals	过氧自由基
17	Polysulfone	聚砜	44	Glass transition temperature	玻璃化转变温度
18	Polyethylene	聚乙烯	45	Thermal-oxidative degradation	热氧降解
19	Polyvinyl chloride	聚氯乙烯	46	Nylon	尼龙
20	Polymethylmethacralate	聚甲基丙烯酸甲酯	47	Microorganism	微生物
21	GPC	凝胶渗透色谱	48	Fungus	真菌
22	Deactivators	去活化剂	49	Bacteria	细菌
23	Polyisobutene	聚异丁烯	50	Rodent	啮齿动物
24	Degree of branching	支化度	51	Moth	飞蛾
25	Low density polyethylene	低密度聚乙烯	52	Bulk polymerization	本体聚合
26	Hyperbranched polymers	超支化聚合物	53	Emulsion polymerization	乳液聚合
27	High density polyethylene	高密度聚乙烯	54	Suspension polymerization	悬浮聚合

Continued table

No.	English	Chinese	No.	English	Chinese
55	Initiator	引发剂	82	Tissue engineering	组织工程
56	Polyolefin	聚烯烃	83	Bioresorbability	生物吸附性
57	Polydispersity	多分散性	84	Biocompatibility	生物相容性
58	Butadiene-styrene resin	丁二烯-苯乙烯树脂	85	Lactobacillus	乳酸菌
59	Gelation	凝胶化	86	In-vivo	体内
60	End group modification	端基修饰	87	Vinyl	乙烯基
61	co-polymerization	共聚	88	Masterbatch	母料
62	blend	共混	89	Weathering	风化作用
63	Trioxane	三聚甲醛	90	Embrittlement	脆化
64	Glycidyl methacrylate	甲基丙烯酸缩水甘油酯	91	Charring	炭化
65	POM	聚甲醛	92	Polyureas	聚脲
66	Ethylene propylene diene monomer	三元乙丙橡胶	93	Permebability	渗透性
67	biphenol-A	双酚A	94	Hydrolases	水解酶类
68	Alcoholysis	醇解	95	Pharamceutical	药剂学
69	Viscous flow temperature	黏流温度	96	Resorbable surgical sutures	可吸收外科缝线
70	Orientation	取向	97	Resorbable orthopedic devices	可吸收矫形装置
71	Semicrystalline	半晶态	98	Solvolysis	溶解度
72	Mulch plastic film	覆膜	99	Glycolysis	糖酵解
73	ABS resin	丙烯腈-二乙烯丁二烯树脂	100	Stereochemical	立体化学
74	Additives	添加剂	101	Depolymerization	解聚
75	Quenchers	淬灭剂	102	Unzipping reaction	开链反应
76	Thermal stabilizer	热稳定剂	103	Chain-end scission	链端断裂
77	Photo stabilizer	光稳定剂	104	Depropagation	链断裂作用
78	Antioxidant	抗氧化剂	105	Random-chain fragmentation	随机链断裂
79	Antiozonant	抗臭氧剂	106	Random chain-breaking reaction	随机链断裂反应
80	Polymer recycling	聚合物回收	107	Oligomer	低聚物
81	Orthopaedics	骨科	108	Polycondensation	缩聚

Appendix I Typical wordlist used in this book

Continued table

No.	English	Chinese	No.	English	Chinese
109	Polyaddition	加聚	139	Glycosidic bond	糖苷键
110	Tetramer	四聚体	140	D-glucose	D-葡萄糖
111	Pentamer	五聚体	141	Aerobic	有氧的
112	Backbone	骨架	142	Anaerobic	厌氧
113	Disproportionation	歧化	143	Amylase	淀粉酶
114	Quaternary carbon radicals	季碳自由基	144	Aspergillus niger	黑曲霉
115	Pyromellitic dianhydride	均苯四甲酸二酐	145	Aflatoxin	黄曲霉毒素
116	homolytic	均裂的	146	Cellulose	纤维素
117	Reductant	还原剂	147	Trichoderma viride	绿色木霉
118	Oxidant	氧化剂	148	Cellobiase	纤维二糖酶
119	Pelleting	造粒	149	Black yeast	黑酵母
120	Radical scavenger	自由基清除剂	150	Vulcanization	硫化
121	Benzofuranone	苯并呋喃酮	151	Exoenzymes	胞外酶
122	Synergetic effect	协同效应	152	Penicillium	青霉
123	Extrusion	挤出	153	Moraxella lacunata	陷窝莫拉菌
124	Antagonistic	拮抗的	154	Papain	木瓜蛋白酶
125	Chain scission	断链	155	Rearrangement	重排
126	Fluorescence	荧光	156	Chemiluminescence	化学发光
127	Photon	光子	157	Peracids	过氧酸
128	Photosensitizer	光敏剂	158	Exoenergetic	放热的
129	Quantum yield	量子产率	159	Volatile	易挥发的
130	Atactic	无规	160	Metal injection molding	金属注射成型
131	Light shielding agent	遮光剂	161	Acetalization	缩醛化
132	UV absorber	紫外线吸收剂	162	Autoacceleration (Gel Effect)	自加速效应（凝胶化效应）
133	Enol	烯醇	163	Auto-accelerared oxidation	自加速氧化
134	Chromophoric groups	发色团	164	Kevlar	凯芙拉纤维
135	Zwitterion	两性离子	165	Macromonomers	大分子单体
136	Metabolization	代谢	166	Ozonides	抗臭氧剂
137	Xenobotics	异种疗法	167	Ozone cracking	臭氧裂解
138	Starch	淀粉	168	Telechelic polymers	遥爪聚合物

Appendix II Chemical structure, name and abbreviation of typical polymers

In this section, a wordlist for polymer degradation is given as following to help the readers to further understand the English edition of "Polymer Degradation".

Chemical structure, name and abbreviation of typical polymers

No.	Chemical structure	Name	Abbreviation
1	$\left[CH_2-CH(CH_3) \right]_n$	Polypropylene	PP
2	$\left[CH_2-CH(C_6H_5) \right]_n$	Polystyrene	PS
3	$\left[CF_2-CF_2 \right]_n$	Poly(tetrafluoroethylene)	PTFE
4	$\left[O-C_6H_4-C(CH_3)_2-C_6H_4-O-C_6H_4-SO_2-C_6H_4 \right]_n$	Polysulfone	PSF
5	$\left[CH_2-CH_2 \right]_n$	Polyethylene	PE
6	$\left[CH_2-CHCl \right]_n$	Polyvinyl chloride	PVC
7	$\left[CH_2-C(CH_3)(COOCH_3) \right]_n$	Polymethylmethacralate	PMMA
8	$\left[CH_2-C(CH_3)_2 \right]_n$	Polyisobutene	PIB
9	$\left[CH_2-O \right]_n$	Polyformaldehyde	POM

Appendix Ⅱ Chemical structure, name and abbreviation of typical polymers

Continued table

No.	Chemical structure	Name	Abbreviation
10		Polycarbonate	PC
11		Polylactic acid	PLA
12		Polybutylene terephthalate	PBT
13		Polyethylene glycol terephthalate	PET
14		Poly(propylene adipate)	N/A
15		Poly(lactic-co-glycolic acid)	PLGA
16		Polycaprolactone	PCL
17		Poly(hydroxybutyrate)	PHB
18		Poly(ethylene terephthalate)	PETG
19		Polymethyl acrylate	PMA
20		Poly(α-methylstyrene)	PAMS

No.	Chemical structure	Name	Abbreviation
21		Poly(methacrylonitrile)	PMAN
22		Polyvinyl alcohol	PVA
23		Poly(*tert*-butylmethacrylate)	N/A
24		Polyacrylonitrile	PAN
25		Polyvinyl acetate	PVAc
26		Poly-*m*-phenylene terephthamide	Nomex
27		Poly aramid	Kevlar
28		Polyamic acid	PAA
29		Polyimide	PI
30		Polybutylene	PB
31		Polyisoprene	PI

Appendix Ⅱ Chemical structure, name and abbreviation of typical polymers

Continued table

No.	Chemical structure	Name	Abbreviation
32		Polybutadiene	PB
33		Polychloropene	N/A
34		Acrylonitrile butadiene styrene plastic	ABS
35		styrenic block copolymers	SBS
36		Polyetherimides	PEI
37		Polyadiohexylenediamine	Nylon-66
38		Poly(hydroxypropylmethylacrylamide)	N/A
39		Polyacrylic acid	PAA
40		Polyaryletheretherketone	PEEK

Appendix III Typical parameters for the characteristic viscosity-molecular weight relationship of typical polymers

Typical parameters for the characteristic viscosity-molecular weight relationship of typical polymers

Polymer	Solvent	$T/°C$	$K/10^3$ (g/mL)	$a/$(g/mL)	Molecular weight $M/10^3$	Method[①]
LDPE	Decalin Xylene	70 105	6.8 1.76	0.675 0.83	<200 11.2–180	O O
HDPE	Chloronaphthalene	125	4.3	0.67	48–950	L
PP	Decalin Tetraline	135 135	1.00 0.80	0.80 0.80	100–1100 40–650	L O
PIB	Cyclohexane	30	2.76	0.69	37.8–700	O
PB	Toluene	30	3.05	0.725	53–490	O
PI	Benzene	25	5.02	0.67	0.4–1,500	O
PS	Benzene	20	1.23	0.72	1.2–540	L, S.D
i-PS	Toluene	25	1.7	0.69	3.3–1,700	L
PVC	Cyclohexanone	25	0.204	0.56	19–150	O
PMMA	Acetone Benzene	20 20	0.55 0.55	0.73 0.76	40–8,000 40–8,000	S.D S.D
PVAC	Butanone	25	4.2	0.62	17–1,200	O, S.D
PVA	Water	30	6.62	0.64	30–120	O
PAN	DMF	25	3.92	0.75	28–1,000	O
Nylon-6	85% formic acid	20	7.5	0.70	4.5–16	E
Nylon-66	90% formic acid	20	11	0.72	6.5–26	E
Celluloseacetate	Acetone	25	1.49	0.82	21–390	O
Nitrocellulose	Acetone	25	2.53	0.795	68–224	O
Ethyl cellulose ethoce	Ethyl acetate	25	1.07	0.89	40–140	O
PDMS	Benzene	20	2.00	0.78	33.9–114	L
POM	DMF	150	4.4	0.66	89–285	L
PC	Dichloromethane THF	20 20	1.11 3.99	0.82 0.70	8–270 8–270	S.D S.D
Natural rubber	Toluene	25	5.02	0.67		
PET	Phenol-CCl$_4$(1:1)	25	2.10	0.82	5–25	E
PEG	Water	30	1.25	0.78	10–100	S.D

[①] E for end group analysis, O for osmotic pressure, L for light scattering, S.D for ultracentrifugal sedimentation.

References

[1] ITO M M, GIBBONS A H, QIN D, et al. Structural colour using organized microfibrillation in glassy polymer films[J]. Nature, 2019, 570: 363 – 367.

[2] HUSSANAIN S M, SHAH S Z H, MEGAT-YUSOFF P S M, et al. Degradation and mechanical performance of fibre-reinforced polymer composites under marine environments: a review of recent advancements [J]. Polymer Degradation and Stability, 2023, 215: 110452.

[3] MIRI V, PERSYN O, LEFEBVRE J M, et al. Effect of water absorp-tion on the plastic deformation behavior of nylon 6[J]. European Polymer Journal, 2009, 45: 757 – 762.

[4] CHEN Y L, SPIERING A J H, KARTHIKEYAN S, et al. Mechanically induced chemiluminescence from polymers incorporating a 1, 2 – dioxetane unit in the main chain[J]. Nature Chemistry, 2012, 4: 559 – 562.

[5] VERT M, LI M S, SPENLEHAUER G, et al. Bioresorbability and biocompatibility of aliphatic polyesters[J]. Journal of Materials Science, 1992, 3: 432 – 446.

[6] VON BURKERSRODA F, SCHEDL L, GOPFERICH A, et al. Why degradable polymers undergo surface erosion or bulk erosion[J]. Biomaterials, 2002, 23: 4221 – 4231.

[7] BAMFORD C H, TRIPPER C F H. Comprehensive Chemical Kinetics[M]. Newyork: Elsevier, 1975.

[8] BRZOZOWSJA-STANUCH A, RABIEJ S, FABIA J, et al. Changes in thermal properties of isotactic polypropylene with different additives during aging process [J]. Polimery, 2014, 59: 302 – 307.